Analytical Groundwater Modeling

This book provides a detailed description of how Python can be used to give insight into the flow of groundwater based on analytic solutions. Starting with simple problems to illustrate the basic principles, complexity is added step by step to show how one-dimensional and two-dimensional models of one or two aquifers can be implemented. Steady and transient flow problems are discussed in confined, semi-confined, and unconfined aquifers that may include wells, rivers, and areal recharge. Special consideration is given to coastal aquifers, including the effect of tides and the simulation of interface flow.

Application of Python allows for compact and readable code, and quick visualization of the solutions. Python scripts are provided to reproduce all results. The scripts are also available online so that they can be altered to meet site-specific conditions. This book is intended both as training material for the next generation of university students and as a useful resource for practitioners. A primer is included for those who are new to Python or as a refresher for existing users.

Mark Bakker is a groundwater engineer and a professor at the faculty of Civil Engineering and Geosciences of the Delft University of Technology. He has taught groundwater and Python classes for more than two decades and is the originator and co-developer of many Python-based open-source modeling tools, including Timml, Ttim, Pastas, and Flopy.

Vincent Post is a hydrogeologist with research interests in coastal hydrogeology and groundwater quality. After a career in academia and research he founded Edinsi Groundwater in 2021. Python has become indispensable in his everyday professional life, and he actively promotes its use in hydrogeology by teaching courses and developing Python-based tools.

Analytical Groundwater Modeling

Theory and Applications using Python

Mark Bakker and Vincent Post

CRC Press
Taylor & Francis Group
Boca Raton London New York

CRC Press is an imprint of the
Taylor & Francis Group, an **informa** business

Cover illustration: Eva van Aalst. https://www.evavanaalst.com/

First published 2022
by CRC Press/Balkema
Schipholweg 107C, 2316 XC Leiden, The Netherlands
e-mail: enquiries@taylorandfrancis.com
www.routledge.com – www.taylorandfrancis.com

CRC Press/Balkema is an imprint of the Taylor & Francis Group, an informa business

Library of Congress Cataloging-in-Publication Data
Names: Bakker, Mark, author. | Post, Vincent, author.
Title: Analytical groundwater modeling : theory and applications using
Python / Mark Bakker, Faculty of Civil Engineering and Geosciences,
Technical University of Delft, Delft, The Netherlands, Vincent Post,
Edinsi Groundwater, Nederhorst den Berg, The Netherlands.
Description: First Edition. | Boca Raton : Taylor and Francis, [2022] |
Includes bibliographical references and index.
Identifiers: LCCN 2021059473 (print) | LCCN 2021059474 (ebook) | ISBN 9781138605633
(Hardback) | ISBN 9781138029392 (Paperback) | ISBN 9781315206134 (eBook)
Subjects: LCSH: Groundwater flow–Computer simulation. | Python (Computer
program language)
Classification: LCC GB1197.7 .B336 2022 (print) | LCC GB1197.7 (ebook) |
DDC 551.4901/13–dc23/eng20220406
LC record available at https://lccn.loc.gov/2021059473
LC ebook record available at https://lccn.loc.gov/2021059474

ISBN: 978-1-138-60563-3 (hbk)
ISBN: 978-1-138-02939-2 (pbk)
ISBN: 978-1-315-20613-4 (ebk)

DOI: 10.1201/9781315206134

Typeset in Times New Roman
by codeMantra

Contents

Preface		ix
About this book		xi
Acknowledgement		xiii
Authors		xv

0   Basics of groundwater flow — 1

0.1	Hydraulic head	1
0.2	Darcy's experiment	2
0.3	Representative elementary volume	2
0.4	Hydraulic conductivity and porosity	3
0.5	Aquifers and aquitards	4
0.6	Storage and transient flow	5
0.7	Darcy's law for groundwater modeling	6
0.8	Dupuit and Forchheimer	7
0.9	Groundwater models	7

1   Steady one-dimensional flow with constant transmissivity — 9

1.1	Flow between two rivers	10
1.2	Areal recharge between two rivers	13
1.3	Areal recharge between an impermeable boundary and a river	18
1.4	Flow through two zones of different transmissivities	22

2   Steady one-dimensional semi-confined flow — 25

2.1	Flow from a canal to a drained area	26
2.2	Flow between a lake and a drained area	29
2.3	Flow to a long river of finite width	32
2.4	Flow to a river in a two-aquifer system	35
2.5	Areal recharge between two rivers in a two-aquifer system	39

3   Steady one-dimensional unconfined flow with variable saturated thickness — 43

3.1	Areal recharge between an impermeable boundary and a river	44
3.2	Flow over a step in the aquifer base	48
3.3	Combined confined/unconfined flow with areal recharge	50

4   Steady one-dimensional flow in coastal aquifers   55

    4.1   Confined interface flow   56
    4.2   Unconfined interface flow   60
    4.3   Combined confined/semi-confined interface flow   64

5   Transient one-dimensional flow   69

    5.1   Step changes in surface water level   70
    5.2   Periodic changes in surface water level   77
    5.3   Areal recharge between two rivers   80
    5.4   Solutions with Laplace transforms   87
    5.5   Unconfined flow with variable transmissivity   91

6   Steady two-dimensional flow to wells   95

    6.1   Radially symmetric flow on a circular island   98
    6.2   Wells near rivers and impermeable boundaries   102
    6.3   Wells near an inhomogeneity boundary   108
    6.4   Wells in a semi-confined aquifer   110
    6.5   Wells in a two-aquifer system   113

7   Steady two-dimensional flow to wells in uniform background flow   117

    7.1   A single well in uniform background flow   117
    7.2   Well capture zones   122
    7.3   A well in uniform background flow near a river   124
    7.4   A well in uniform background flow near a river with a leaky stream bed   129
    7.5   A well in uniform background flow near the coast   134

8   Analytic element modeling of steady two-dimensional flow   137

    8.1   Uniform flow and wells   137
    8.2   Line-sinks for modeling rivers and streams   144
    8.3   Area-sinks for modeling areal recharge   148

9   Transient two-dimensional flow   151

    9.1   Wells in confined and unconfined aquifers   152
    9.2   Wells with a periodic discharge   158
    9.3   Wells in a semi-confined aquifer   160
    9.4   Wells with wellbore storage and skin effect   164
    9.5   Wells in a two-aquifer system   167

10 Steady two-dimensional flow in the vertical plane   171

    10.1 Vertical anisotropy   172
    10.2 Flow to a partially penetrating stream   174
    10.3 Flow over a step in the base   181
    10.4 Spatially varying head at the top of the aquifer   185
    10.5 Interface flow towards the coast   190
    10.6 Interface flow below a strip island   194

A Python primer   199

   A.1 Basics   200
   A.2 Loops and if statements   203
   A.3 The numpy package and arrays   205
   A.4 The matplotlib package for visualization   210
   A.5 Functions   213
   A.6 The scipy package for scientific computing   216

Numerical answers to selected problems   219
Bibliography   221
Index   225

Preface

The importance of groundwater as a water supply source can hardly be overstated. Groundwater exists everywhere in the subsurface. Extraction of water from aquifers supports livelihoods, agriculture, and a variety of economic activities worldwide. Groundwater is an integral part of the hydrological cycle where it sustains ecosystems, such as wetlands and estuaries, and provides the base flow of streams and rivers of all sizes. The flow of water through the subsurface constitutes a major transport pathway for dissolved substances and plays a key role in the planetary cycle of many chemical elements.

Groundwater resources have come under enormous stress in many parts of the world. Concerns over the quantity and quality of groundwater are the result of past and ongoing mismanagement. Overexploitation has resulted in significant lowering of water levels. Unsustainable pumping in coastal areas has resulted in seawater intrusion. And careless disposal and handling of chemicals at the land surface has resulted in contamination that threatens precious resources. Addressing these problems is possible only with adequate understanding of the flow in the subsurface.

The investigation of groundwater systems relies first and foremost on measurements. Due to the difficulty and costs involved in obtaining these, data is commonly available at a few points only, which severely limits the knowledge of the three-dimensional system under investigation. It is sometimes possible to predict future developments based on measured time series, but it is rarely possible to make predictions from measured time series for new interventions in the system or when anthropogenic, environmental, or climatic changes alter the system's response.

Groundwater models are essential to gain understanding of the behavior of groundwater systems and to evaluate management scenarios that cannot be tested at full scale in the field. Groundwater models, like all models, are simplified versions of a complex reality. It is an important skill for every modeler to decide how a system can be simplified without losing the essence. It is unfortunately often appealing to add complexity to models because it exists, even though it doesn't affect the outcome significantly.

The essence of a groundwater flow problem is captured in a set of equations called the mathematical model. Simple models can often be solved analytically. Before the advent of grid-based methods, groundwater solutions were necessarily simplified until they could be solved, for example by deriving an analytic solution to the mathematical model. Analytic solutions give insight into the flow of groundwater that is more difficult to obtain with grid-based solutions. It is often forgotten that analytic solutions have benefited from the advent of computer technology, just like grid-based methods. Especially the development of high-level programming languages makes it easy to evaluate analytic solutions and visualize their results.

Models play an important role in solving the formidable water problems of our planet, as they are used as instruments to better understand, manage, and protect our water resources. The authors hope that the knowledge contained in this book contributes to better groundwater models and, ultimately, better-informed management decisions in the future.

About this book

This book provides a fresh treatment of the fundamentals of groundwater flow and is intended as both training material for the next generation of university students and a useful resource for practitioners who want to re-familiarize themselves with the basics while simultaneously learning about the capabilities of Python. Programming skills were part of groundwater modeling in the early days of the personal computer age. Classic books from, e.g., Bear and Verruijt (1987) and Strack (1989) already included computer code (on paper or floppy disks), but that was before high-level computer languages such as Python conquered the world. The example programs developed in this book may serve as templates that can be modified and extended to simulate settings or processes not covered in this book. The focus is on the fundamentals of the quantitative methods used to simulate, understand, and solve groundwater problems.

The premise of this book is that any groundwater modeler needs a fundamental understanding of the theory on which mathematical models of groundwater flow are based. Even when a modeler primarily makes grid-based models, it is important to understand the main approximations that form the basis of the mathematical model, their range of applicability, and the main consequences for the flow field. Insight in fundamental flow systems forms the backbone of any good modeler. This book provides detailed descriptions of analytic solutions of many fundamental groundwater problems. Mathematical derivations are included for the simpler problems. The solutions of more complicated problems are presented without derivation, but their validity is verified.

The structure of this book is such that the theoretical treatments of the modeling methods are exemplified by short pieces of Python computer code that demonstrate the implementation of the theory. Python is an open-source programming language that has rapidly gained popularity within science and engineering and is likely going to remain one of the most popular programming languages over the next decades. Its simple yet powerful syntax and interactive workflow make learning and using Python a breeze. The development of powerful packages for scientific computing and visualization have made it easier than ever to evaluate, explore, and visualize both simple and complicated groundwater solutions. Python is a great language to write scripts to evaluate groundwater models. Scripts may be combined with output and explanations (including mathematical equations) in Jupyter Notebooks, resulting in executable documents that are a full, transparent and reproducible record of a model or analysis.

All code examples presented in this book are available as Jupyter Notebooks from `www.github.com/pythongroundwaterbook`, which facilitates easy exploration of the examples in this book. Even though the notebooks save the reader from having to do a lot of typing, there are important benefits to entering the code examples from the printed book into one's own notebook or Python interpreter. As one types, one takes much better notice of the subtle details contained in the code fragments. Especially when errors occur, one is forced to analyze their cause, which may take some time but at the reward of a deeper understanding and a more lasting learning experience. Therefore, especially novice students are encouraged to use this method and not immediately trust that they understand the worked examples in the notebooks. Hands-on exercises are provided and commonly require short Python scripts to be solved. Numerical answers to selected exercises are included at the end of the book.

This book starts with a brief introduction of the principles of groundwater flow. This chapter is assigned number 0 to remind the reader that Python uses a zero-based numbering system. All other numbering in this book starts at 1. The remainder of this book consists of two parts, each consisting of five chapters. The first part concerns one-dimensional flow while the second part concerns two-dimensional flow. A variety of flow systems are discussed, including confined, unconfined, semi-confined, and multi-aquifer flow. Both steady and transient flow are considered. Most examples include evaluation of both the head and the flow, sometimes including travel times. Special attention is paid to interface flow in coastal aquifers.

For each solution, at least three code blocks are presented. The first code block contains the parameter values used in the example. The second code block contains the solution. And the third code block contains a few basic commands to produce visual output, which resemble the presented figures. The visual output appears right below the third code block, or on the next page if the output does not fit on the page. The exact code to produce the final figures included in this book is not provided as it contains some detailed (and repetitive) commands to make the figures look nice for the book. For those groundwater engineers and hydrogeologists that are not familiar with Python programming, or need a refresher of some of the programming concepts, a short primer on Python programming is included in the Appendix of this book.

Acknowledgement

The authors wish to thank Erik Anderson and Michael Teubner for reading and commenting on a draft of the entire book. The groundwater theory presented in this book is based on the principles of groundwater flow that were taught by Prof. Arnold Verruijt at the Delft University of Technology and Prof. Otto Strack at the University of Minnesota. Lastly, the authors thank the developers of Python, the `numpy-scipy-matplotlib` stack (Harris et al., 2020; Virtanen et al., 2020; Hunter, 2007), and Project Jupyter (Kluyver et al., 2016) for the development of a remarkable, invaluable, and enjoyable programming language and ecosystem for exploratory computing.

Authors

Mark Bakker is a groundwater engineer and a professor at the faculty of Civil Engineering and Geosciences of the Delft University of Technology. He has taught groundwater and Python classes for more than two decades and is the originator and co-developer of many Python-based open-source modeling tools, including Timml, Ttim, Pastas, and Flopy.

Vincent Post is a hydrogeologist with research interests in coastal hydrogeology and groundwater quality. After a career in academia and research he founded Edinsi Groundwater in 2021. Python has become indispensable in his everyday professional life, and he actively promotes its use in hydrogeology by teaching courses and developing Python-based tools.

Chapter 0

Basics of groundwater flow

Groundwater is defined as the water beneath the water table, where the pores are fully saturated with water. The flow of groundwater forms the subsurface part of the hydrological cycle. The main source of groundwater is precipitation (rain or snow) that reaches the water table after infiltrating into the soil. The resulting increase in groundwater volume is called recharge, sometimes called meteoric recharge to emphasize that the water originates from atmospheric precipitation.

Natural groundwater discharge occurs where groundwater reaches the land surface, for example in springs, swamps, and wetlands, or directly into rivers, lakes, seas, or oceans. Shallow groundwater can also be extracted by vegetation or through evaporation directly at the soil surface. In many inhabited areas, significant amounts of water are extracted through pumping wells.

0.1 Hydraulic head

Groundwater may be reached by digging or drilling a hole in the ground. The level to which water rises in a borehole is called the hydraulic head, or simply the head. A piezometer is an observation well specifically designed to measure the head and consists of a pipe with a slotted screen at the bottom. The water level in a piezometer rises to a height such that the weight of the water column in the piezometer equals the pressure of the groundwater directly adjacent to the well screen. In a well-constructed piezometer (i.e., one that has no leaks or internal flow), the pressure distribution with depth is hydrostatic

$$p = \rho g(h - z) \tag{0.1}$$

where p [M/L/T^2] is the pressure with respect to the atmosphere in the groundwater at the elevation of the well screen, ρ [M/L^3] is the (mean) density of the water in the piezometer, g [L/T^2] is the gravitational acceleration, h [L] is the head, and z [L] is the elevation of the well screen (positive upward) with respect to a reference level. Equation (0.1) can be rewritten as (Figure 0.1)

$$h = z + \frac{p}{\rho g} \tag{0.2}$$

The term z is also referred to as the elevation head, and the term $p/(\rho g)$ as the pressure head. The pressure p equals zero at the water table so that the hydraulic head is equal to the elevation head, $h = z$. Comparing water levels from different wells requires all water levels to be expressed with respect to a common reference level or datum (e.g., mean sea level).

Exercise 0.1: Consider a piezometer of which the top is 20 m above the datum. The well screen is 30 m below the top of the piezometer. A pressure sensor is lowered into the piezometer to a

DOI: 10.1201/9781315206134-0

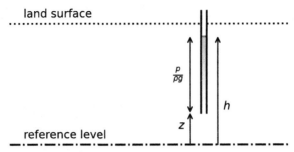

Figure 0.1 Definition of the head h and its components according to Eq. (0.2).

depth of 10 m below the top of the piezometer and records an absolute pressure of $p_{abs} = 180$ kPa. The piezometer is filled with freshwater with a density $\rho = 1000$ kg/m^3. The atmospheric pressure is measured to be 101 kPa. Use $g = 9.8$ m/s^2. Compute the head in the well.

0.2 Darcy's experiment

In 1856, the French engineer Henry Darcy published the results of a series of experiments on the flow of water through a column packed with different kinds of sand (Darcy, 1856). The experiments consisted of a soil column of length L and cross-sectional area A (Figure 0.2). Darcy measured the discharge Q through the column and the difference Δh in the head of the water between two measurement ports.

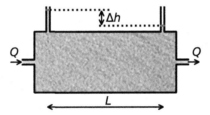

Figure 0.2 Schematic representation of Darcy's experiment. Darcy used a vertical column, but a horizontal column works just as well.

The experiments revealed that there is a linear relationship between the discharge Q through the soil column and the hydraulic head drop Δh over distance L, which can be expressed as

$$Q = kA\frac{\Delta h}{L} \tag{0.3}$$

The proportionality constant k was introduced because the relationship between the discharge and the head gradient differed between types of sand. Nowadays, k is known as the hydraulic conductivity, which has the dimensions L/T. Darcy's law is not a physical law, but an empirical relation. It became known as a law because it performs remarkably well over a large range of flow conditions.

0.3 Representative elementary volume

Water particles move through small openings in the subsurface, along convoluted pathways caused by the irregular shape of the pore space. The simulation of groundwater flow at this microscopic scale requires knowledge of the three-dimensional shape of the entire pore space, which is not feasible. A different approach is required for larger spatial scales, called the macroscopic scale. At the macroscopic scale, the net movement of groundwater through the subsurface is approximated using parameters that are a representative average of the behavior at the microscopic scale. This is what Darcy established empirically: The friction and viscous forces that work against the flow of water are captured by the hydraulic conductivity k. This parameter thus represents the effective action of the processes at the microscopic scale, without detailed knowledge of the pore geometry at this scale.

The variables for the macroscopic description of groundwater flow are associated with a spatial scale that is representative of the system considered. The volume for which a macroscopic property is valid is called the representative elementary volume (REV). The REV must not be too small, because it will not form a representative average. In the extreme, an REV smaller than the size of a pore or soil particle either represents an opening between soil particles or a soil particle itself, but not the average of many of these combined. The size of an REV must therefore exceed a multiple of the representative pore size. Conversely, an REV must not be too large either as it may no longer be suited to resolve the spatial variability of the subsurface at the macroscopic level. If the properties of the subsurface vary over a few meters, the REV must be significantly smaller than a few meters to capture this variability in a model. The REV relevant to practical groundwater problems varies. For unconsolidated sands, with a pore and grain size on the order of tenths of a millimeter, an REV of a few centimeters already captures the average behavior.

0.4   Hydraulic conductivity and porosity

The hydraulic conductivity k is a measure of how easy a fluid can flow through the subsurface. The hydraulic conductivity is not just a property of the subsurface, but also a property of the fluid that flows through the subsurface and may be written as

$$k = \frac{\kappa \rho g}{\mu} \tag{0.4}$$

where κ [L^2] is the intrinsic permeability that depends on the porous medium, while the density, ρ, and the dynamic viscosity, μ [M/L/T], are properties of the fluid that flows through the subsurface. This means that warmer water will have a higher hydraulic conductivity than colder water, even when it flows through the same subsurface, as the viscosity of warmer water is smaller than the viscosity of colder water. Typical values of the hydraulic conductivity of water through different materials are presented in Table 0.1.

Table 0.1   Hydraulic conductivity ranges of different materials

Material	k (m/d)
Clay	< 0.0001
Sandy clay	0.0001 − 0.001
Peat	0.0001 − 0.01
Silt	0.001 − 0.01
Very fine sand	0.1 − 1
Fine sand	1 − 10
Coarse sand	10 − 100
Sand with gravel	100 − 1000
Gravel	> 1000

Source: Modified from Verruijt (1970).

The porosity, n, of the subsurface is defined as the ratio of the volume of pores, V_p, to the total volume, V_t.

$$n = \frac{V_p}{V_t} \tag{0.5}$$

The porosity does not appear in the equation for the hydraulic conductivity (Eq. 0.4), but the intrinsic permeability κ depends on the characteristics of the subsurface, including the pore volume, the pore size distribution, the connectivity between pores, the shape of the pores, and the roughness of the subsurface material. The porosity therefore partially determines the intrinsic permeability κ and thus indirectly influences k as well.

Exercise 0.2: Compute the hydraulic conductivity (in m/d) of freshwater flowing through sand in winter when the infiltrating water has a temperature of 5°C and in summer when the infiltrating water has a temperature of 20°C. Given: $\kappa = 2 \cdot 10^{-11}$ m^2, $\mu = 1.5 \cdot 10^{-3}$ Ns/m^2 at 5°C, and $\mu = 1 \cdot 10^{-3}$ Ns/m^2 at 20°C.

0.5   Aquifers and aquitards

The subsurface consists of layers of varying permeability. The permeable layers are referred to as aquifers. Sediments such as sand or gravel tend to form good aquifers, as their pores are relatively wide and well connected, so that water can flow with relative ease. Layers with a lower permeability are called aquitards, leaky layers, or semi-confining layers. Truly impermeable layers do not exist in nature, but if the hydraulic conductivity is so low that flow through them is negligible, they are sometimes called aquicludes or confining layers.

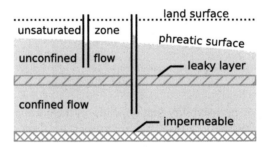

Figure 0.3 Schematic cross section of the subsurface showing aquifers and aquitards.

An aquifer is unconfined when the saturated domain is bounded at the top by the water table, also called the phreatic surface (Figure 0.3). Confined aquifers are separated from other aquifers, or the land surface, by an aquitard. Flow is referred to as confined when the head is above the top of the aquifer (Figure 0.3). The head in confined aquifers can drop below the top of the aquifer, resulting in unconfined flow (see Section 3.3). Artesian conditions exist when the head in a confined aquifer is higher than the elevation of the land surface, which means that groundwater flows freely to the surface if a well is installed, without the need for a pump.

In systems consisting of multiple aquifers separated by aquitards, groundwater may move from one aquifer to another through aquitards. This is referred to as multi-aquifer or inter-aquifer flow. Flow in aquitards can often be approximated as vertical. The larger the thickness of an aquitard, the more difficult it is for water to flow across it. The ratio of the thickness of an aquitard to its vertical hydraulic conductivity is called the hydraulic resistance and is a measure of the (im)permeability of a leaky layer. Flow in aquifers tends to be predominantly horizontal. The larger the thickness of an aquifer, the more water it can transmit. The ability of an aquifer to transmit water in the horizontal direction is called the transmissivity, which is the product of the hydraulic conductivity and the aquifer thickness.

A groundwater modeler must always decide about the base of the groundwater flow system, which is commonly formed by a low-permeable layer that leaks so little water that it can be neglected for the problem of interest. While it is common to refer to the bottom of the system as the impermeable base, it must be realized that there is always some exchange of water across the bottom boundary of the system considered. Ideally, a confining unit can be defined that serves as the base of the model, and in this case, a better term for the base of the model is the basal aquitard. For very thick aquifer systems (like in sedimentary basins and deltas, which can be many kilometers deep), this is not always feasible, and a pragmatic choice must be made.

0.6   Storage and transient flow

When there are no significant changes in head (and thus flow) through time, the flow is said to be steady. Analyses of steady flow are appropriate when long-term, average conditions are of interest, for example for the delineation of wellhead protection areas for water supply wells, or when an end situation is of interest, as for the dewatering of an aquifer for a construction project. When the heads do not change with time, the storage volume remains constant in time, which implies that the inflow and outflow of water are equal in the system considered.

In reality, groundwater flow is always transient. Transient flow occurs because boundary conditions and system stresses change through time, for example changing aquifer recharge, changing river levels, or varying pumping rates of wells. A specific application of transient flow analysis is the evaluation of aquifer properties by induced stresses, such as pumping tests.

The storage in an unconfined aquifer increases when the water table rises. As an example, consider a column within an unconfined aquifer with constant horizontal cross-sectional surface area A. When the water table is raised Δh, the volume of water in the column increases by an amount

$$\Delta V = SA\Delta h \tag{0.6}$$

where S [-] is the storativity or storage coefficient of the unconfined aquifer. When the porous medium above the water table is completely dry before the rise, the storativity of the unconfined aquifer is equal to the porosity. In reality, the storativity is always smaller than the porosity, as there is always soil water present above the water table. The storativity of an unconfined aquifer is referred to as the phreatic storage coefficient or the more cryptic specific yield and is represented by the variable S_p. Typical values of the phreatic storage for different materials are presented in Table 0.2.

Storage in a confined aquifer is physically more complicated. In a confined aquifer, all pores are entirely filled with water. Additional water can nevertheless be stored through expansion of the pore space and, albeit only to a small extent, compression of the water. When the head

Table 0.2 Phreatic storage of different materials

Material	S_p (−)
Clay	0.01 − 0.05
Silt	0.03 − 0.19
Sand	0.1 − 0.35
Gravel	0.14 − 0.35

Source: Modified from Fitts (2013).

rises by Δh, the volume of water increases by an amount ΔV according to Eq. (0.6), but the storage coefficient is much smaller than that of unconfined aquifers. For most unconsolidated aquifers, the ability of the aquifer to expand is significantly larger than the ability of the water to compress. The storage coefficient of a confined aquifer is a function of the aquifer thickness: The thicker the aquifer, the more water can be stored per unit surface area A, and thus the greater the storage coefficient. The storage coefficient of a confined aquifer is

$$S = S_s H \tag{0.7}$$

where S_s [1/L] is the specific storage of the aquifer (the storage per unit aquifer thickness). Typical values of the specific storage are presented in Table 0.3.

Table 0.3 Specific storage values of different materials

Material	S_s (m^{-1})
Clay	$1 \cdot 10^{-3} - 2 \cdot 10^{-2}$
Sand	$1 \cdot 10^{-5} - 1 \cdot 10^{-3}$

Source: Modified from Anderson et al. (2015).

0.7 Darcy's law for groundwater modeling

The basic form of Darcy's law (Eq. 0.3) relates the total flow Q through a column of soil to the head gradient. For groundwater modeling, it is more useful to consider the discharge per unit area

$$q = \frac{Q}{A_q} = k\frac{\Delta h}{L} \tag{0.8}$$

where A_q is the cross-sectional area perpendicular to the flow. The dimensions of the specific discharge vector are L/T. It is important to note, however, that the specific discharge is not a velocity (although it is sometimes confusingly referred to as the Darcy velocity), but a discharge per unit area.

Groundwater flow can be in any direction and is expressed in terms of the specific discharge vector $\vec{q} = (q_x, q_y, q_z)$ with components in each of the three coordinate directions. The general form of Darcy's law relates the components of the specific discharge vector to the head gradient in each direction

$$q_x = -k\frac{\partial h}{\partial x} \qquad q_y = -k\frac{\partial h}{\partial y} \qquad q_z = -k\frac{\partial h}{\partial z} \tag{0.9}$$

In Eq. (0.9), the hydraulic conductivity is the same for flow in all directions. In some aquifers, the hydraulic conductivity is anisotropic and must be represented by a tensor. When the principal directions of the hydraulic conductivity tensor coincide with the coordinate directions, Darcy's law may be written as

$$q_x = -k_x\frac{\partial h}{\partial x} \qquad q_y = -k_y\frac{\partial h}{\partial y} \qquad q_z = -k_z\frac{\partial h}{\partial z} \tag{0.10}$$

where k_x, k_y, and k_z are the hydraulic conductivity in the three coordinate directions (the principal components of the hydraulic conductivity tensor).

Water in the subsurface can only flow through the pores. The average velocity vector $\vec{v}$ can be obtained from the specific discharge vector through division by the porosity n

$$\vec{v} = \frac{\vec{q}}{n} \tag{0.11}$$

The vector $\vec{v}$ is the average velocity of the groundwater because it describes the flow at the macroscopic level, not the flow of individual water particles at the microscopic level of the pores.

Exercise 0.3: Compute the average groundwater velocity for sand with $k = 20$ m/d, a porosity of $n = 0.3$, and a head that drops 2 m every 1000 m. Compute v in both meters per day and meters per month.

0.8   Dupuit and Forchheimer

All groundwater flow is three-dimensional, but for modeling purposes, flow may be approximated as one-dimensional or two-dimensional. At the regional scale, most aquifers are relatively thin in comparison with their areal extent and are therefore referred to as shallow aquifers. Flow is predominantly horizontal in such aquifers and may be approximated as two-dimensional in the horizontal plane. The amount of horizontal flow is described by the discharge vector $Q = (Q_x, Q_y)$. The discharge vector is the total flow in the aquifer integrated over the saturated thickness H of the aquifer. In other words, it is the discharge per unit width of aquifer. The components of the discharge vector may be computed as

$$Q_x = Hq_x = -kH\frac{\partial h}{\partial x} \qquad Q_y = Hq_y = -kH\frac{\partial h}{\partial y} \tag{0.12}$$

where the hydraulic conductivity is horizontally isotropic ($k_x = k_y = k$). The dimensions of the discharge vector are L^2/T. This approximation was first proposed by Dupuit (1863) and Forchheimer (1886), so it is referred to as Dupuit flow or Dupuit–Forchheimer flow.

The Dupuit–Forchheimer approximation essentially means that the pressure distribution is hydrostatic in an aquifer. This is very convenient, because that means that the head h does not depend on the vertical coordinate in an aquifer. The Dupuit–Forchheimer approximation does not mean, however, that flow is horizontal. Water can still flow vertically, but the resistance to vertical flow is neglected within an aquifer; a detailed analysis is given in Chapter 1. The accuracy of the Dupuit–Forchheimer approximation is further explored in Chapter 10.

0.9   Groundwater models

Groundwater models are needed to gain an understanding of how groundwater flows and to try to make predictions, because most phenomena or proposed interventions cannot be tested at full scale in the field. Groundwater models are mathematical solutions to the differential equation that describes flow in an aquifer system. The differential equation is obtained through the combination of a mass balance equation and Darcy's law, yielding a differential equation in terms of the hydraulic head. A mathematical model of a groundwater problem consists of a differential equation and a set of mathematical boundary conditions, which are derived from boundaries and conditions of the physical groundwater system. In this book, the mathematical problem is solved analytically.

In each of the upcoming chapters, analytical solutions are presented for a wide range of flow problems. Chapters 1–5 deal with flow systems that may be approximated as one-dimensional in the horizontal direction. This is followed by solutions to two-dimensional flow systems in the horizontal plane in Chapters 6–9, including flow systems that are radially symmetric (e.g., flow near a pumping well). Finally, Chapter 10 deals with two-dimensional flow in the vertical plane, including a comparison with one-dimensional Dupuit–Forchheimer solutions.

Chapter 1

Steady one-dimensional flow with constant transmissivity

The study of one-dimensional groundwater flow problems gives basic insights that are quickly lost in the analysis of two-dimensional or three-dimensional flow fields. Groundwater flow fields can be approximated as one-dimensional if the flow in one direction dominates the flow in the other directions, e.g., near coast lines, lake shores, rivers, or drainage ditches. Especially at the regional scale, the flow tends to be essentially horizontal because the length of the flow paths in the horizontal direction (on the order of kilometers) is much larger than the water movement in the vertical direction (on the order of tens of meters or less).

The steady mass balance states that

$$\text{Mass in} - \text{Mass out} = 0 \tag{1.1}$$

The mass balance of a fluid reduces to a volume balance when the density of the fluid is approximated as constant. The steady volume balance is

$$\text{Volume in} - \text{Volume out} = 0 \tag{1.2}$$

The volume balance for steady one-dimensional groundwater flow in an aquifer is derived by considering a small section of the aquifer that is Δx long and $\Delta y = 1$ wide in the direction normal to the plane of flow. The inflow consists of horizontal flow from the left, $Q_x(x)$, and recharge at the top, N [L/T]. The outflow consists of horizontal flow at the right, $Q_x(x + \Delta x)$ (see Figure 1.1). The discharge vector Q_x is the discharge per unit width of the aquifer, so the dimensions are L^2/T, while the recharge is the discharge per unit area with dimensions L/T.

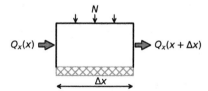

Figure 1.1 Steady water balance in a section of an aquifer that is Δx long and $\Delta y = 1$ wide in the direction normal to the plane of flow.

The volume balance is written for the time period Δt. Substitution of the appropriate volumes in the volume balance equation (1.2) gives

$$Q_x(x)\Delta t + N\Delta x\Delta t - Q_x(x + \Delta x)\Delta t = 0 \tag{1.3}$$

Division by Δx and Δt and rearrangement of terms gives

$$\frac{Q_x(x + \Delta x) - Q_x(x)}{\Delta x} = N \tag{1.4}$$

DOI: 10.1201/9781315206134-1

In the limit for $\Delta x \to 0$, this gives the differential form of the continuity equation for steady one-dimensional horizontal flow

$$\frac{\mathrm{d}Q_x}{\mathrm{d}x} = N \qquad (1.5)$$

The term $\mathrm{d}Q_x/\mathrm{d}x$ is also known in mathematics as the one-dimensional divergence of the discharge vector. The divergence is the mathematical equivalent of outflow minus inflow of a flow field.

The transmissivity of an aquifer is approximated as constant in all but the last solution in this chapter. An aquifer with a constant transmissivity may be a confined aquifer with horizontal top and bottom and constant hydraulic conductivity k. Or it can be an unconfined aquifer where the saturated thickness is approximated as constant. Either way, the Dupuit–Forchheimer approximation is adopted so that the head is approximated as constant in the vertical direction. The horizontal flow is distributed equally over the aquifer thickness H so that the discharge vector Q_x, the integrated specific discharge over the thickness of the aquifer, may be written as

$$Q_x = Hq_x = -T\frac{\mathrm{d}h}{\mathrm{d}x} \qquad (1.6)$$

where $T = kH$ is the transmissivity of the aquifer and q_x is given by Darcy's law (Eq. 0.9). The continuity equation for this problem is (1.5). Substitution of Eq. (1.6) for Q_x in Eq. (1.5) gives

$$\frac{\mathrm{d}}{\mathrm{d}x}\left(T\frac{\mathrm{d}h}{\mathrm{d}x}\right) = -N \qquad (1.7)$$

Because the transmissivity T is assumed constant, the differential equation simplifies to

$$\frac{\mathrm{d}^2h}{\mathrm{d}x^2} = -\frac{N}{T} \qquad (1.8)$$

This is a second-order linear ordinary differential equation known as the Poisson equation. The general solution is obtained by integrating Eq. (1.8) twice, which gives, for the case that N is constant,

$$h = -\frac{N}{2T}x^2 + Ax + B \qquad (1.9)$$

where A and B are integration constants that need to be determined from boundary conditions.

1.1   Flow between two rivers

The first problem that is solved is for confined flow between two long and parallel rivers that fully penetrate the aquifer (Figure 1.2). The rivers are a distance L apart, and there is no areal recharge ($N = 0$). The rivers are in direct contact with the aquifer so that the head in the aquifer at the river is equal to the head in the river. The head at $x = 0$ is h_0, and the head at $x = L$ is h_L.

$$h|_{x=0} = h_0 \qquad (1.10)$$
$$h|_{x=L} = h_L \qquad (1.11)$$

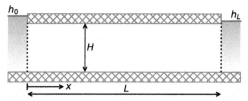

Figure 1.2 Flow between two rivers.

Application of the boundary conditions to the general solution (1.9) gives

$$A = \frac{h_L - h_0}{L} \qquad (1.12)$$

$$B = h_0 \qquad (1.13)$$

so that the solution for the head becomes a straight line

$$h = \frac{(h_L - h_0)x}{L} + h_0 \qquad (1.14)$$

The discharge vector Q_x is obtained by differentiating Eq. (1.14) with respect to x and inserting the result into Eq. (1.6)

$$Q_x = -T\frac{\mathrm{d}h}{\mathrm{d}x} = -T\frac{h_L - h_0}{L} \qquad (1.15)$$

Note that Q_x is constant throughout the flow field.

The solution is computed below and plotted in Figure 1.3. Note that the porosity n and elevation z_b of the bottom of the aquifer are defined but not used here. They will be used when visualizing the flow field and computing the travel time in the following.

```
# parameters
L = 1000 # aquifer length, m
H = 10 # aquifer thickness, m
zb = -6 # aquifer bottom, m
k = 10 # hydraulic conductivity, m/d
n = 0.3 # porosity, -
T = k * H # transmissivity, m^2/d
h0 = 6 # specified head at the left boundary, m
hL = 4 # specified head at the right boundary, m
```

```
# solution
x = np.linspace(0, L, 100)
h = (hL - h0) * x / L + h0
Qx = - T * (hL - h0) / L * np.ones_like(x)
```

```
# basic plot
plt.subplot(121)
plt.plot(x, h)
plt.subplot(122)
plt.plot(x, Qx);
```

The flow field in a vertical cross section may be visualized using the streamplot function of matplotlib, which takes as input a two-dimensional grid of points, and for each grid

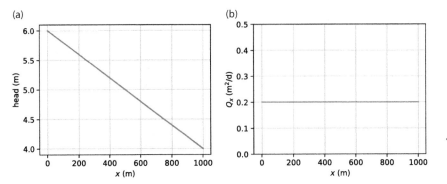

Figure 1.3 Flow between two rivers with $N = 0$; head (left) and Q_x (right).

point, the horizontal and vertical components of the specific discharge vector. The horizontal component of the specific discharge vector q_x is computed from the discharge vector Q_x as

$$q_x = \frac{Q_x}{H} = -k\frac{h_L - h_0}{L} \tag{1.16}$$

and does not vary in the vertical direction. The vertical component of the specific discharge vector q_z equals zero. The streamplot function draws lines in the direction of the flow and adds arrows to indicate the direction of flow. For this case with constant flow, a small grid of 2 by 2 is sufficient. The vertical exaggeration is set with the aspect keyword of the subplot function. The vertical exaggeration is noted as "VE" in the label of the vertical axis (Figure 1.4).

```
# solution
xg, zg = np.meshgrid(np.linspace(0, L, 2), np.linspace(zb, zb + H, 2))
qx = -k * (hL - h0) / L * np.ones_like(xg)
qz = np.zeros_like(xg)
```

```
# basic streamplot
plt.subplot(111, aspect=25)
plt.streamplot(xg, zg, qx, qz, color='C1', density=0.2);
```

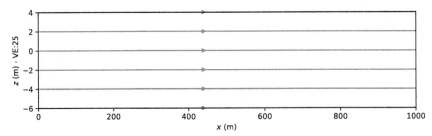

Figure 1.4 Streamplot in a vertical cross section for flow between two rivers with $N = 0$.

The (average) velocity in the horizontal direction v_x is obtained through division of q_x by the porosity n (obviously, the velocity in the vertical direction v_z is zero). For this simple problem, the (average) travel time from the left river to the right river can be computed as L/v_x. More formally, the travel time can be obtained from the definition of the horizontal velocity vector

$$v_x = \frac{q_x}{n} = \frac{\mathrm{d}x}{\mathrm{d}t} \tag{1.17}$$

Integration gives the travel time t_{tr}

$$\int_0^{t_{\text{tr}}} dt = \int_0^L \frac{dx}{v_x} = \int_0^L -\frac{Ln}{k(h_L - h_0)} dx \tag{1.18}$$

which gives

$$t_{\text{tr}} = \frac{L^2 n}{k(h_0 - h_L)} = \frac{L}{v_x} \tag{1.19}$$

```
# travel time
vx = -k * (hL - h0) / (n * L)
print(f'mean velocity: {vx:.4f} m/d')
print(f'travel time from left river to right river: {L / vx:.0f} days')
```

```
mean velocity: 0.0667 m/d
travel time from left river to right river: 15000 days
```

1.2 Areal recharge between two rivers

The transmissivity of an unconfined aquifer may be approximated as constant and equal to kH if the variation of the saturated thickness, caused by the change in water table elevation, is small compared to the total saturated thickness. In the following example, an unconfined aquifer is bounded on the left and right sides by two long parallel rivers that fully penetrate the aquifer (Figure 1.5). The rivers are a distance L apart and in direct contact with the aquifer so that

$$h|_{x=0} = h_0 \tag{1.20}$$
$$h|_{x=L} = h_L \tag{1.21}$$

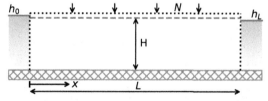

Figure 1.5 Areal recharge between two rivers.

Because T is approximated as constant, the general equation (1.8) is still applicable and the integration constants for $N > 0$ are

$$A = \frac{h_L - h_0}{L} + \frac{N}{2T}L \tag{1.22}$$
$$B = h_0 \tag{1.23}$$

so that the solution becomes

$$h = -\frac{N}{2T}(x^2 - Lx) + \frac{(h_L - h_0)x}{L} + h_0 \tag{1.24}$$

The discharge vector Q_x is obtained by differentiation of Eq. (1.24) in combination with Eq. (1.6).

$$Q_x = -T\frac{dh}{dx} = N\left(x - \frac{L}{2}\right) - T\frac{h_L - h_0}{L} \tag{1.25}$$

The solution is computed below.

```
# parameters
L = 1000 # aquifer length, m
H = 10 # saturated thickness, m
zb = -5 # aquifer bottom, m
k = 10 # hydraulic conductivity, m/d
n = 0.3 # porosity, -
T = k * H # transmissivity, m^2/d
h0 = 6 # specified head at the left boundary, m
hL = 4 # specified head at the right boundary, m
N = 0.001  # areal recharge, m/d
```

```
# solution
x = np.linspace(0, L, 100)
h = -N / (2 * T) * (x ** 2 - L * x) + (hL - h0) * x / L + h0
Qx = N * (x - L / 2) - T * (hL - h0) / L
```

```
# basic plot
plt.subplot(121)
plt.plot(x, h)
plt.subplot(122)
plt.plot(x, Qx);
```

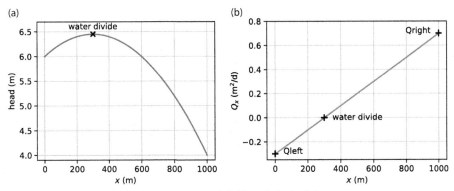

Figure 1.6 Areal recharge between two rivers; head (left) and Q_x (right).

Note that there is a groundwater divide between the two rivers (Figure 1.6). Part of the recharge flows to the left river, and part of the recharge flows to the right river. The discharge into the left river ($x = 0$) and the right river ($x = L$) are computed below. The flow into the left river plus the flow into the right river is exactly equal to the total recharge: $NL = 1$ m^2/d. At the water divide, the head gradient equals zero so that $Q_x = 0$.

```
print(f'discharge into left river: {-Qx[0]:.3f} m^2/d')
print(f'discharge into right river: {Qx[-1]:.3f} m^2/d')
```

```
discharge into left river: 0.300 m^2/d
discharge into right river: 0.700 m^2/d
```

Exercise 1.1: Compute the location of the water divide for the parameters that are provided above. Verify that the location of the water divide is consistent with the discharge figures computed above.

Exercise 1.2: Determine what the water level in the left river must be so that exactly all water that infiltrates flows to the right river, while no water from the left river flows to the right river; use the parameters of the example above. Plot h and Q_x vs. x.

Exercise 1.3: Consider the case that $h_0 = h_1 = 10$ m. The head halfway between the two rivers is measured to be 9 m. Compute the areal recharge for this case using the parameters of the example above. Plot h vs. x.

In this solution, the flow is approximated as one-dimensional, and both h and Q_x are functions of x only. In a vertical cross section, however, the flow must be two-dimensional. After all, water infiltrates the aquifer at the top and discharges into the rivers to the left and right (see Figure 1.5). This leads to an apparent contradiction: How can water flow vertically in a one-dimensional solution?

To resolve this contradiction, consider Darcy's law for the vertical component of the specific discharge

$$q_z = -k_z \frac{\partial h}{\partial z} \tag{1.26}$$

where k_z is the hydraulic conductivity in the vertical direction. As the Dupuit–Forchheimer approximation is adopted, the head h is a function of x only, so $\partial h / \partial z = 0$. This means that q_z is also equal to zero, except for the case that the vertical hydraulic conductivity k_z is infinitely large; the product of a term that goes to zero and a term that goes to infinity can result in a finite number (under certain mathematical conditions). Although this may seem strange at first (the vertical hydraulic conductivity is, of course, a finite number), this result can be interpreted to mean that the resistance to vertical flow is neglected (e.g., Strack, 1984). In other words, the energy it takes to flow vertically within an aquifer can be neglected with respect to the energy it takes to flow horizontally. This is often a good approximation, as groundwater commonly travels large distances horizontally (order of kilometers), but only small distances vertically (order of meters).

The vertical component of the flow cannot be computed from Darcy's law (as $k_z = \infty$ causes all kinds of troubles), but can be obtained from continuity of flow. A differential equation for continuity of flow in the vertical x, z plane may be derived by considering a vertical aquifer section that is Δx long and Δz high. The fluxes on the four sides are shown in Figure 1.7. Application of the volume balance for steady flow (In − Out = 0) gives

$$[q_x(x + \Delta x, z) - q_x(x, z)]\Delta z \Delta t + [q_z(x, z + \Delta z) - q_z(x, z)]\Delta x \Delta t = 0 \tag{1.27}$$

Division by $\Delta x \Delta z \Delta t$ and taking the limit for Δx and Δz going to zero gives

$$\frac{\partial q_x}{\partial x} + \frac{\partial q_z}{\partial z} = 0 \tag{1.28}$$

In mathematical terms, the divergence of the specific discharge vector equals zero for steady flow in the vertical plane.

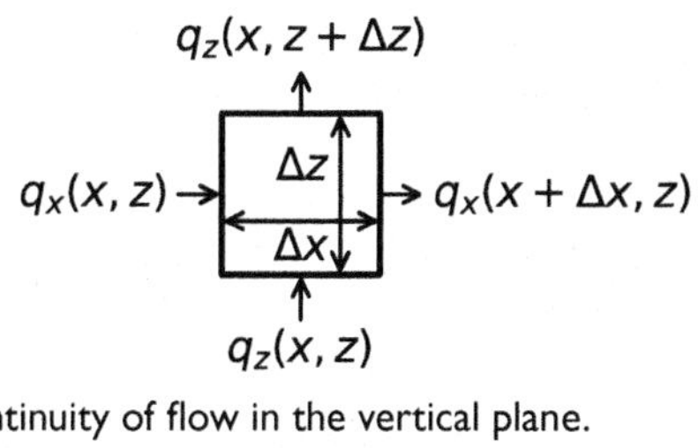

Figure 1.7 Two-dimensional continuity of flow in the vertical plane.

The horizontal component q_x is obtained from the discharge vector (Eq. 1.25) as

$$q_x = \frac{Q_x}{H} = -k\frac{dh}{dx} = \frac{N}{H}\left(x - \frac{L}{2} - \frac{T}{N}\frac{h_L - h_0}{L}\right) \tag{1.29}$$

Differentiation of q_x and substitution of the result in Eq. (1.28) gives

$$\frac{\partial q_z}{\partial z} = -\frac{N}{H} \qquad (1.30)$$

which can be integrated to give

$$q_z = -\frac{Nz}{H} + C \qquad (1.31)$$

The constant of integration is obtained from the condition that at the impermeable base of the aquifer ($z = z_b$), the vertical component of flow is zero so that $C = Nz_b/H$ and

$$q_z = -\frac{N(z - z_b)}{H} \qquad (1.32)$$

Note that the boundary condition at the top of the aquifer $q_z|_{z=z_b+H} = -N$ is also satisfied, and that q_z decreases with depth according to a simple linear function. Expressions (1.29) and (1.32) are used below to create a streamplot for approximate two-dimensional flow in the vertical plane (Figure 1.8).

```
# solution
xg, zg = np.meshgrid(np.linspace(0, L, 10), np.linspace(zb, zb + H, 5))
qx = (N * (xg - L / 2) - T * (hL - h0) / L) / H
qz = - N * (zg - zb) / H
```

```
# basic streamplot
plt.subplot(111, aspect=25)
plt.streamplot(xg, zg, qx, qz, color='C1');
```

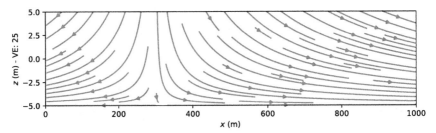

Figure 1.8 Streamplot in a vertical cross section for areal recharge between two rivers.

A second method to visualize the two-dimensional flow field in a vertical cross section is to compute and contour the stream function. The stream function is a function that is constant along streamlines. The difference of the stream function values between two points is equal to the amount of water flowing between the two points, and the value of the stream function increases to the right when looking in the direction of flow.

The stream function in a vertical cross section is denoted with the small Greek letter psi, ψ [L^2/T]. Here, ψ is arbitrarily set to zero along the impermeable base of the aquifer, as the impermeable base is a streamline. As a result, the stream function value along the top of the aquifer is equal to $\psi = -Q_x$ so that the difference between the stream function values along the bottom and top of the aquifer is always equal to the discharge Q_x in the aquifer. The stream function value along the top of the aquifer is $-Q_x$ and not Q_x, because of the convention that the stream function increases to the right when looking in the direction of flow.

A contour plot of the stream function is shown in Figure 1.9. It is created by defining an array psi of the stream function. The stream function is computed at a number of points along the bottom and top of the aquifer. The contouring routine interpolates linearly between the stream function computed along the top and bottom of the aquifer. This is exactly correct for this problem, as this means that the discharge is divided equally over the vertical. The amount of flow between two adjacent streamlines is always the same when a constant contouring interval is used (this is not the case for the lines obtained with the streamplot function used previously), so when two streamlines are closer together, water flows faster. In this case, the flow between two adjacent streamlines is $NL/20$, as 20 contour intervals are plotted. The streamlines are evenly spaced every 50 m at the top of the aquifer, as the inflow is equal to $q_z = -N$ there and $L = 1000$ m. The vertical streamline that goes through the water divide is the dividing streamline and corresponds to a stream function value $\psi = 0$ (as this streamline continues along the bottom of the aquifer). The outflow to the right river is more than twice the outflow to the left river. Note that, indeed, the stream function increases to the right when looking in the direction of flow. The inflow along the top of the aquifer is uniform and equal to N. The outflow into the rivers is also uniform in the vertical direction. This means, for example, that water that enters the aquifer halfway between the water divide and the river flows into the river at an elevation exactly equal to the middle of the aquifer.

```
# solution
x = np.linspace(0, L, 100)
Qx = N * (x - L / 2) - T * (hL - h0) / L
psi = np.zeros((2, len(x)))
psi[1, :] = -Qx # at the top of the aquifer, the stream function is -Qx
xg = x
zg = [zb, zb + H]
```

```
# basic streamline plot
plt.subplot(111, aspect=25)
cs = plt.contour(xg, zg, psi, 20, colors='C1', linestyles='-')
plt.clabel(cs, fmt='%1.2f');
```

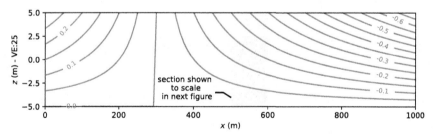

Figure 1.9 Streamlines in a vertical cross section for areal recharge between two rivers. Vertical exaggeration (VE) is 25.

In the plot of the streamlines, the scales along the horizontal and vertical directions are not the same: The horizontal distance is actually 25 times larger than the vertical distance. When the scales along both axes are set equal, it is not possible to see the flow lines anymore for the entire cross section. For illustration purposes, the streamlines are plotted to scale (i.e., no vertical exaggeration) for the section from $x = 500$ to $x = 600$ (colored gray in Figure 1.9) in the figure below, showing that the streamlines are indeed predominantly horizontal (Figure 1.10).

```
# basic streamline plot of part of flow field without vertical exaggeration
plt.subplot(111, aspect=1, xlim=(500, 600))
cs = plt.contour(xg, zg, psi, 20, colors='C1', linestyles='-');
```

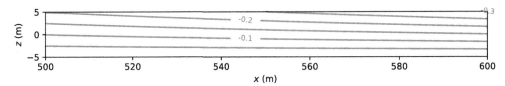

Figure 1.10   Streamlines in a vertical cross section for the gray area of cross section shown in Figure 1.9 (no vertical exaggeration).

Exercise 1.4: Make a contour plot of the stream function for the case that the head in the left river is so high that exactly all water flows to the right river (i.e., the water divide is at the left river).

It is common practice to draw contours of the head and streamlines on the same plot, as they form what is called a flow net where the streamlines cross the head contours at right angles, as will be demonstrated for two-dimensional solutions in later chapters. This is only the case when the hydraulic conductivity is isotropic. Here, the solution for flow in the vertical plane is based on the Dupuit–Forchheimer approximation, which is equivalent to the case where the vertical hydraulic conductivity is much larger than the horizontal hydraulic conductivity so that the head contours and streamlines don't form a flow net.

Exercise 1.5: Draw a flow net showing head contours and streamlines for the example on the same graph.

The travel time for water that enters at the top of the aquifer and leaves the aquifer at the left or right river can again be computed from the horizontal component of the velocity vector $v_x = q_x/n$ (since v_x is not a function of the vertical coordinate), as was done in the previous section. This gives the travel time $t_{\mathrm{tr}}(x_0, x_1)$ from location $x = x_0$ to location $x = x_1$

$$t_{\mathrm{tr}}(x_0, x_1) = \int_{x_0}^{x_1} \frac{\mathrm{d}x}{v_x} = \frac{nH}{N} \ln\left[\frac{v_x|_{x=x_1}}{v_x|_{x=x_0}}\right] \tag{1.33}$$

Similarly, the travel time from the top of the aquifer $z_b + H$ to elevation z is obtained from the vertical component of the velocity vector $v_z = q_z/n$, which gives

$$t_{\mathrm{tr}}(z) = \int_{z_b+H}^{z} \frac{\mathrm{d}z}{v_z} = \frac{nH}{N} \ln\left(\frac{H}{z - z_b}\right) \tag{1.34}$$

The travel time from the top of the aquifer to elevation z is only a function of z. That is an interesting result, famously obtained by Vogel (1967), because it means that all water at a certain depth has the same age for the case of uniform recharge.

Exercise 1.6: Compute how long it takes for water that enters the aquifer at the top of the aquifer halfway between the water divide and the right river to flow to the right river. Use the parameters of the example and a porosity $n = 0.3$. Compute the travel time in two ways: using the equation for the travel time from x_0 to x_1, and the equation for the travel time from the top of the aquifer to the halfway thickness of the aquifer. They should, of course, give the same answer.

1.3   Areal recharge between an impermeable boundary and a river

The problem of the previous section is modified by replacing the river on the left side by an impermeable boundary (see Figure 1.11). The two boundary conditions become

$$\left.\frac{\mathrm{d}h}{\mathrm{d}x}\right|_{x=0} = 0 \tag{1.35}$$

$$h|_{x=L} = h_L \tag{1.36}$$

The boundary condition at $x = L$ is referred to as a fixed-head or Dirichlet boundary condition. The boundary condition at $x = 0$ is referred to as a specified-flux or Neumann boundary condition. The Dirichlet and Neumann boundary conditions are also referred to as type 1 and type 2 boundary conditions, respectively. In physical terms, the impermeable boundary can represent a water divide or a rock mass with very low permeability.

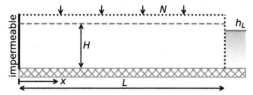

Figure 1.11   Areal recharge between an impermeable wall and a river.

The general solution to the differential equation is again Eq. (1.9), of which the derivative with respect to x is

$$\frac{dh}{dx} = -\frac{N}{T}x + A \tag{1.37}$$

Application of boundary condition Eq. (1.35) at $x = 0$ gives

$$A = 0 \tag{1.38}$$

so that application of boundary condition (1.36) leads to the following solution

$$h = -\frac{N}{2T}(x^2 - L^2) + h_L \tag{1.39}$$

and

$$Q_x = -T\frac{dh}{dx} = Nx \tag{1.40}$$

All the recharge flows to the river on the right, so the outflow into the right river is NL. An example is provided below (Figure 1.12). Note that the line for the head function is horizontal at the left boundary, indicating the no-flow condition. The saturated thickness is approximated as constant and equal to $H = 10$ m in the example below, but the distance between the head and the bottom of the aquifer ($z_b = -5$ m) varies from 14 m on the left side to 9 m on the right side. The question can be raised whether it is reasonable to approximate the saturated thickness as constant if it varies by this much. This question is addressed in Chapter 3.

```
# parameters
L = 1000 # aquifer length, m
H = 10 # approximate saturated aquifer thickness, m
zb = -5 # aquifer bottom, m
k = 10 # hydraulic conductivity, m/d
T = k * H # transmissivity, m^2/d
hL = 4 # specified head at the right boundary, m
N = 0.001  # areal recharge, m/d
```

```
# solution
x = np.linspace(0, L, 100)
h = -N / (2 * T) * (x ** 2 - L ** 2) + hL
Qx = N * x
```

```
# basic plot
plt.subplot(121)
plt.plot(x, h)
plt.subplot(122)
plt.plot(x, Qx);
```

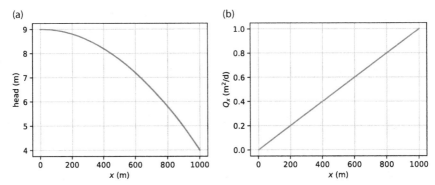

Figure 1.12 Areal recharge between an impermeable boundary and a river; head (left) and Q_x (right).

Exercise 1.7: Compute how long it takes for water that infiltrates at $x = L/2$ to flow to the river. Use $n = 0.3$.

Exercise 1.8: Derive and plot a solution for the head for the case that $h|_{x=0} = h_0$ and $\frac{dh}{dx}|_{x=L} = 0$, where $h_0 = 6$ m.

Next, the boundary condition on the right side is modified by adding a layer with a resistance between the aquifer and the river (Figure 1.13). This resistance represents the low-permeable sediments that typically make up the river bed. In reality, the inflow occurs along the wet perimeter of the river, but to simplify the mathematical treatment, the leaky river bed is approximated here as a vertical section between the aquifer and the river with width w_h and horizontal hydraulic conductivity k_h. The specific discharge through the layer is computed with Darcy's law as

$$q_x = -k_h \frac{h_L - h_R}{w_h} = \frac{h_L - h_R}{c} \tag{1.41}$$

where h_L is the head in the aquifer at $x = L$, h_R is the water level in the river, and c is referred to as the hydraulic resistance of the river bed and is defined as

$$c = \frac{w_h}{k_h} \tag{1.42}$$

Note that c has the dimension time. The discharge vector into the river is obtained through multiplication of q_x by the saturated thickness in the leaky layer, which is approximated here as H.

$$Q_x|_{x=L} = H \frac{h_L - h_R}{c} = C(h_L - h_R) \tag{1.43}$$

where $C = H/c$ is called the river bed conductance. This boundary condition represents a relation between the head and the flow and is referred to as a Robin boundary condition or a type 3 boundary condition.

For this modified problem, the flow into the river is known upfront from continuity of flow and is equal to the total recharge NL so that

$$Q_x|_{x=L} = C(h_L - h_R) = NL \tag{1.44}$$

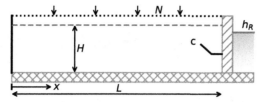

Figure 1.13 Areal recharge between an impermeable boundary and a river with river bed resistance.

and the head in the aquifer at $x = L$ is

$$h_L = \frac{NL}{C} + h_R \tag{1.45}$$

The solution for the head and flux can now be obtained from Eqs. (1.39) and (1.40). The shape of the head curve is the same as for the case without entry resistance (Figure 1.12), but all heads are raised by an amount NL/C (Figure 1.14), because all the water needs to flow through the resistance layer into the river. For the parameters of the previous example, but with an entry resistance of $c = 10$ d, the head increase is 1 m.

```
# additional parameters
hR = 4 # water level in river, m
c = 10 # hydraulic resistance of river bed, d
```

```
# solution
x = np.linspace(0, L, 100)
C = H / c
hL = N * L / C + hR
h = -N / (2 * T) * (x ** 2 - L ** 2) + hL
Qx = N * x
```

```
# basic plot
plt.subplot(121)
plt.plot(x, h)
plt.subplot(122)
plt.plot(x, Qx);
```

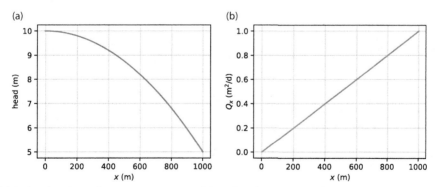

Figure 1.14 Areal recharge between an impermeable wall and a river with riverbed conductance; head (left) and Q_x (right).

Exercise 1.9: Compute the head in the aquifer at $x = L$ for the case with river bed conductance, but now the boundary condition on the left side is $h|_{x=0} = 6$ m. Plot the head vs. x.

1.4   Flow through two zones of different transmissivities

In the previous solutions, the hydraulic conductivity and saturated thickness were approximated as constant. Here, the aquifer consists of two zones of different, but constant transmissivities (Figure 1.15). The one-dimensional confined flow is between two parallel rivers with specified heads. The rivers are a distance L apart. The zone on the left has length L_0, hydraulic conductivity k_0, and thickness H_0. The zone on the right has length L_1, hydraulic conductivity k_1, and thickness H_1. Hence, the transmissivity in the left zone is $T_0 = k_0 H_0$ and the transmissivity in the right zone is $T_1 = k_1 H_1$. Recall that the general solution for the head in a confined aquifer (Eq. 1.9) was derived for a constant transmissivity. The solution is valid in zone 0 and in zone 1, but with different integration constants. For $N = 0$, the solution in zone 0 is

$$h = A_0 x + B_0 \qquad \text{zone 0} \tag{1.46}$$

and in zone 1

$$h = A_1 x + B_1 \qquad \text{zone 1} \tag{1.47}$$

where A_0, A_1, B_0, and B_1 are constants to be determined from boundary conditions.

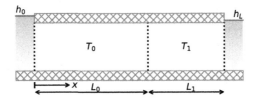

Figure 1.15 Confined flow through two zones of different transmissivities.

There are two boundary conditions and two continuity conditions, because the head and flow are continuous across the boundary between zones 0 and 1

$$h|_{x=0} = h_0 \tag{1.48}$$
$$h|_{x=L} = h_L \tag{1.49}$$
$$h|_{x=L_0^-} = h|_{x=L_0^+} \tag{1.50}$$
$$Q_x|_{x=L_0^-} = Q_x|_{x=L_0^+} \tag{1.51}$$

where the superscripts $^+$ and $^-$ stand for evaluation just to the left and just to the right of $x = L_0$, respectively. Application of these four conditions and a bit of algebra gives the head solution

$$h = \frac{T_1 L}{T_0 L_1 + T_1 L_0} \frac{(h_L - h_0)}{L} x + h_0 \qquad \text{zone 0} \tag{1.52}$$

$$h = \frac{T_0 L}{T_0 L_1 + T_1 L_0} \frac{(h_L - h_0)}{L}(x - L) + h_L \qquad \text{zone 1} \tag{1.53}$$

and the flow solution

$$Q_x = -\frac{T_0 T_1 L}{T_0 L_1 + T_1 L_0} \frac{(h_L - h_0)}{L} \qquad \text{zones 0 and 1} \tag{1.54}$$

The discharge vector is once again constant for this solution (because $N = 0$), so the amount of water that flows into the aquifer on the left must discharge into the river on the right.

The head function consists of two straight segments (1.16). The head gradient is smaller in the zone with the larger transmissivity and the head gradient is larger in the zone with the smaller transmissivity such that the product of the transmissivity and the head gradient is equal for both zones. In the example below, the aquifer is divided in two halves ($L_0 = L_1$) and the transmissivity of the left half of the aquifer is 10 times larger than the transmissivity of the right half of the aquifer. As a result, the head gradient is 10 times smaller in the left zone as compared to the right zone.

```
# parameters
h0 = 6 # specified head at the left boundary, m
hL = 4 # specified head at the right boundary, m
L0 = 500 # length of zone 0, m
L1 = 500 # length of zone 1, m
T0 = 200 # transmissivity of zone 0, m^2/d
T1 = 20 # transmissivity of zone 1, m^2/d
L = L0 + L1 # total length, m
```

```
# solution
x = np.linspace(0, L, 101)
h = np.empty_like(x)
h[x < L0] = T1 * L / (T0 * L1 + T1 * L0) * (hL - h0) * x[x < L0] / L + h0
h[x >= L0] = T0 * L / (T0 * L1 + T1 * L0) * (hL - h0) * (x[x >= L0] - L) / L + hL
Qx = -T0 * T1 * L / (T0 * L1 + T1 * L0) * (hL - h0) / L
hhalfway = h[50]
print(f'head halfway: {hhalfway:.2f} m')
print(f'head gradient zone 0: {(hhalfway - h0) / L0: .6f}')
print(f'head gradient zone 1: {(hL - hhalfway) / L1: .6f}')
```

```
head halfway: 5.82 m
head gradient zone 0: -0.000364
head gradient zone 1: -0.003636
```

```
# basic plot
plt.subplot(121)
plt.plot(x, h)
plt.subplot(122)
plt.plot([0, L], [Qx, Qx]);
```

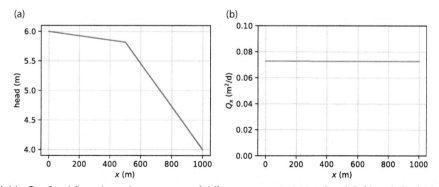

Figure 1.16 Confined flow through two zones of different transmissivities; head (left) and Q_x (right).

The solution for Q_x (Eq. 1.54) is compared with the solution for a homogeneous aquifer with transmissivity T (Eq. 1.15), which shows that the transmissivity of the two zones may be replaced by a single effective transmissivity T_{eff} for the whole aquifer that results in the correct flow rate

$$T_{\text{eff}} = \frac{T_0 T_1 L}{T_0 L_1 + T_1 L_0} = \frac{L}{L_1/T_1 + L_0/T_0} \qquad (1.55)$$

The effective transmissivity is equal to the harmonic mean of the transmissivities of each zone, weighted by the zone lengths. The harmonic mean is commonly (much) smaller than the arithmetic mean.

```
print(f'arithmetic mean transmissivity: {(T0 + T1) / 2: .1f} m^2/d')
print(f'effective transmissivity: {L / (L1 / T1 + L0 / T0): .1f} m^2/d')
```

```
arithmetic mean transmissivity:  110.0 m^2/d
effective transmissivity:   36.4 m^2/d
```

Exercise 1.10: Compute the travel time from the left river to $x = L_0$ and from $x = L_0$ to $x = L$ using that the thickness of both zones is 10 m and the porosity of both zones is 0.3. Repeat the computation for the case that $H_0 = 10$ m and $H_1 = 5$ m, while still using the same value of T_1.

Exercise 1.11: Consider the same case as the example, but now the left boundary is impermeable, and there is a uniform areal recharge of $N = 0.001$ m/d over the entire aquifer. $T_1 = 100$ m^2/d, and all other values are as in the example. First, derive an equation for the head between $x = L_0$ and $x = L$ and compute the head at $x = L_0$. Then derive an equation for the head between $x = 0$ and $x = L_0$. Plot the head vs. x.

Chapter 2

Steady one-dimensional semi-confined flow

In semi-confined aquifers, the aquifer is capped by a leaky layer consisting of a low-permeable material such as peat, shale, or clay. This is in contrast to the confined aquifers considered in the previous chapter, which are capped by a layer that may be approximated as impermeable. The low-permeable layer is referred to as a semi-confining layer or a leaky layer. In what follows, the head above the leaky layer is fixed to a constant level h^* (Figure 2.1). This can correspond to the head in an aquifer above the semi-confining layer, or the water level of a surface water body. The fixed head h^* can also represent the mean water table in a drained area, where the water table is fixed by means of ditches or tile drains; such drained areas are sometimes referred to as polders. The fixed head h^* in the drained area can be in the low-permeable top layer (Figure 2.1) or in a permeable layer separated from the aquifer by a leaky layer (Figure 2.2).

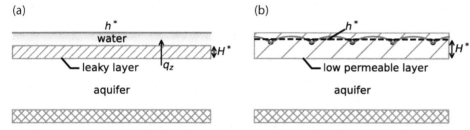

Figure 2.1 Semi-confined aquifer below surface water (left) and semi-confined aquifer below a drained area with a fixed water table in the low-permeable layer (right).

The specific discharge through the leaky layer is approximated as vertical and may be computed with Darcy's law (Eq. 0.9) as

$$q_z = -k^* \frac{\partial h}{\partial z} \approx -k^* \frac{h^* - h}{H^*} \tag{2.1}$$

where h is the head in the aquifer, h^* is the specified head at the top of the leaky layer, k^* is the vertical hydraulic conductivity of the leaky layer, and H^* is either the thickness of the leaky layer (left situation of Figure 2.1) or the distance between the drainage level and the bottom of the low-permeable layer (right situation of Figure 2.1).

The specific discharge through the leaky layer is positive for upward flow (in the positive z direction) and may be written in the form

$$q_z = \frac{h - h^*}{c} \tag{2.2}$$

where c [T] is the resistance to vertical flow of the leaky layer defined as

$$c = H^*/k^* \tag{2.3}$$

DOI: 10.1201/9781315206134-2

The flow in a semi-confined aquifer is governed by the Poisson equation (1.8), where the infiltration rate N is a function of the head difference across the semi-confining layer. Substitution of $N = -q_z$ in Eq. (1.8) gives

$$\frac{\mathrm{d}^2 h}{\mathrm{d}x^2} = \frac{h - h^*}{cT} \tag{2.4}$$

For the case that h^* is a constant, this equation can be rewritten as

$$\frac{\mathrm{d}^2 (h - h^*)}{\mathrm{d}x^2} = \frac{h - h^*}{\lambda^2} \tag{2.5}$$

where

$$\lambda = \sqrt{cT} \tag{2.6}$$

is referred to as the leakage factor [L] and is a characteristic length of the system, as will be explained later.

Differential equation (2.5) is a second-order linear ordinary differential equation known as the modified Helmholtz equation. The general solution is

$$h - h^* = A e^{-x/\lambda} + B e^{x/\lambda} \tag{2.7}$$

where A and B are constants to be determined from boundary conditions. The general solution for the discharge vector is

$$Q_x = \frac{AT}{\lambda} e^{-x/\lambda} - \frac{BT}{\lambda} e^{x/\lambda} \tag{2.8}$$

Alternatively, the general solution may be written as

$$h - h^* = A \sinh(x/\lambda) + B \cosh(x/\lambda) \tag{2.9}$$

with the corresponding discharge vector

$$Q_x = -\frac{AT}{\lambda} \cosh(x/\lambda) - \frac{BT}{\lambda} \sinh(x/\lambda) \tag{2.10}$$

It depends on the problem whether the formulation in terms of exponential functions is more convenient, or the formulation in terms of hyperbolic functions.

2.1   Flow from a canal to a drained area

Consider one-dimensional flow from a canal into a polder. The aquifer is semi-confined and semi-infinite and is bounded on the left by a fully penetrating canal with water level h_0, as shown in Figure 2.2; the head in the polder is fixed to h^*. The governing differential equation is the modified Helmholtz equation, as derived above. The boundary conditions are

$$\begin{aligned} h|_{x=0} &= h_0 \\ h|_{x \to \infty} &= h^* \end{aligned} \tag{2.11}$$

Application of the second boundary condition gives $B = 0$, after which application of the first boundary condition gives $A = h_0 - h^*$ so that the solution for h becomes

$$h = h^* + (h_0 - h^*) e^{-x/\lambda} \tag{2.12}$$

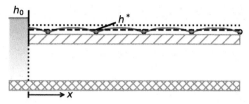

Figure 2.2 Semi-confined flow from a canal to a polder.

Water flows from the canal into the semi-confined aquifer and then through the leaky layer into the overlying polder (provided that the head in the canal is larger than the head in the polder). The discharge vector in the aquifer is

$$Q_x = -T\frac{dh}{dx} = \frac{T(h_0 - h^*)}{\lambda}e^{-x/\lambda} \tag{2.13}$$

An example solution is shown below (Figure 2.3). Equations (2.12) and (2.13) for h and Q_x are implemented in Python functions so that they can easily be reused later on in this section.

```python
# parameters
hstar = 0 # head above leaky layer, m
h0 = 1 # specified head at the left boundary, m
k = 10 # hydraulic conductivity, m/d
H = 40 # aquifer thickness, m
c = 100 # resistance of leaky layer, d
T = k * H # transmissivity, m^2/d
lab = np.sqrt(c * T) # leakage factor, m
print(f'the leakage factor is: {lab:.0f} m')
```

```
the leakage factor is: 200 m
```

```python
# solution
def head(x):
    return (h0 - hstar) * np.exp(-x / lab) + hstar

def disvec(x):
    return k * H * (h0 - hstar) / lab * np.exp(-x / lab)

x = np.linspace(0, 5 * lab, 100)
h = head(x)
Qx = disvec(x)
```

```python
# basic plot
plt.subplot(121)
plt.plot(x, h, 'C0')
plt.subplot(122)
plt.plot(x, Qx, 'C1');
```

The total discharge per meter length of canal normal to the plane of flow is called Q_0 and is obtained by evaluating Q_x at $x = 0$, which gives $Q_0 = T(h_0 - h^*)/\lambda$. The discharge vector Q_x in the aquifer decreases with distance from the canal, as groundwater discharges across the leaky layer into the drainage system of the polder. At a distance $x = \lambda$, Q_x has reduced to $e^{-1}Q_0 = 0.37Q_0$, and at a distance $x = 3\lambda$, Q_x has reduced to only $e^{-3}Q_0 = 0.05Q_0$. Hence, 95% of the water that flows from the canal into the aquifer has discharged into the polder within a distance of 3λ from the canal. That is why the leakage factor λ is a characteristic

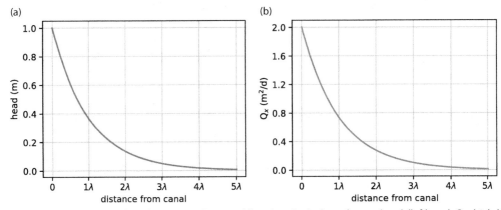

Figure 2.3 Semi-confined flow from a canal to a polder; λ is the leakage factor; head (left) and Q_x (right).

length of the system. Similarly, the difference between the head in the aquifer and the head in the polder reduces from $h_0 - h^*$ at the canal ($x = 0$) to only $0.05(h_0 - h^*)$ at $x = 3\lambda$. Roughly speaking, three times the leakage factor away from the canal, the head in the aquifer is approximately equal to the head h^* above the semi-confined aquifer.

Several checks may be conducted to verify that the solution and its Python implementation (the functions head and disvec) are correct. First, it is verified that the solution meets the boundary conditions in Eq. (2.11).

```
# Verify boundary conditions
print(f'head at x=0: {head(0):.0f} m')
print(f'head at x=infinity: {head(np.inf):.0f} m')
```

```
head at x=0: 1 m
head at x=infinity: 0 m
```

Second, it is verified that the solution meets the governing differential equation (2.5) using an accurate numerical approximation of the second derivative

$$\frac{\mathrm{d}^2 h}{\mathrm{d}x^2} \approx \frac{h(x - \Delta x) - 2h(x) + h(x + \Delta x)}{(\Delta x)^2} \tag{2.14}$$

where Δx is a suitably small increment.

```
# Verify differential equation
x = 200 # example point where deq is checked
delx = 0.001
left_side = (head(x - delx) - 2 * head(x) + head(x + delx)) / delx ** 2
right_side = (head(x) - hstar) / lab ** 2
print(f'left_side of deq : {left_side}')
print(f'right_side of deq: {right_side}')
```

```
left_side of deq : 9.196976513692334e-06
right_side of deq: 9.196986029286058e-06
```

Third and final, it is checked that the specific discharge is indeed $Q_x = -T\mathrm{d}h/\mathrm{d}x$ using the numerical derivative

$$\frac{\mathrm{d}h}{\mathrm{d}x} \approx \frac{h(x + \Delta x) - h(x - \Delta x)}{2\Delta x} \tag{2.15}$$

```
# Verify Qx
Qx = disvec(x)
Qxnum = -T * (head(x + delx) - head(x - delx)) / (2 * delx)
print(f'Qx exact                  : {Qx} m^2/d')
print(f'Qx numerical derivative: {Qxnum} m^2/d')
```

```
Qx exact                  : 0.7357588823428847 m^2/d
Qx numerical derivative: 0.7357588823531103 m^2/d
```

It is important to perform the above three checks to verify a solution. If the solution is correct (and correctly programmed), the solution and the approximation using numerical derivatives must be approximately equal. The number of significant digits that match depends on the size of the increment Δx and on machine accuracy.

Exercise 2.1: Consider the case of the example, but this time the resistance c is unknown. The head in the aquifer is measured in an observation well at $x = 100$ m, where the head is 0.5 m. Compute the resistance of the leaky layer and the leakage factor.

Exercise 2.2: Verify that the total flux from the canal into the aquifer (Q_0) is equal to the integrated leakage through the leaky layer by analytic integration of q_z.

2.2   Flow between a lake and a drained area

One-dimensional flow is modeled between a lake and a drained area, as shown in Figure 2.4. Both the lake and drained area are separated from the aquifer by a semi-confining layer, which has a resistance c_0 below the lake and a resistance c_1 below the drained area. The lake and drained area are separated by a dike of negligible width (the vertical black line at $x = 0$ in the figure). The water level in the lake is h_0^*, while the water level in the drained area is h_1^*. The semi-confined aquifer extends to infinity on both sides. The transmissivity of the aquifer is constant and equal to T.

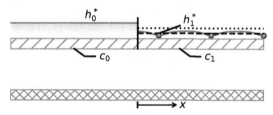

Figure 2.4 Semi-confined flow from a canal to a polder.

Separate differential equations are written for the part of the aquifer below the lake and the part below the drained area

$$\frac{\mathrm{d}^2(h - h_0^*)}{\mathrm{d}x^2} = \frac{h - h_0^*}{\lambda_0^2} \qquad x \leq 0$$

$$\frac{\mathrm{d}^2(h - h_1^*)}{\mathrm{d}x^2} = \frac{h - h_1^*}{\lambda_1^2} \qquad x \geq 0 \tag{2.16}$$

where $\lambda_0 = \sqrt{c_0 T}$ and $\lambda_1 = \sqrt{c_1 T}$.

The boundary conditions are that the head approaches h_0^* far to the left and h_1^* far to the right

$$h|_{x \to -\infty} = h_0^* \qquad h|_{x \to +\infty} = h_1^* \tag{2.17}$$

and continuity requires that the head and flow are continuous across $x = 0$ (see also the conditions at $x = L_0$ in Section 1.4)

$$h_{x=0^-} = h_{x=0^+} \qquad Q_x|_{x=0^-} = Q_x|_{x=0^+} \tag{2.18}$$

where the superscripts $^-$ and $^+$ stand for evaluation just to the left and just to the right of $x = 0$, respectively.

The solutions for the head below the lake and drained area are written as

$$
\begin{aligned}
h &= h_0^* + A_0 e^{-x/\lambda_0} + B_0 e^{x/\lambda_0} \qquad x \le 0 \\
h &= h_1^* + A_1 e^{-x/\lambda_1} + B_1 e^{x/\lambda_1} \qquad x \ge 0
\end{aligned}
\tag{2.19}
$$

Application of the first two boundary conditions gives $A_0 = 0$ and $B_1 = 0$, after which application of the two continuity conditions gives the following two equations

$$
\begin{aligned}
h_0^* + B_0 &= h_1^* + A_1 \\
-T\frac{B_0}{\lambda_0} &= T\frac{A_1}{\lambda_1}
\end{aligned}
\tag{2.20}
$$

Solving for the two remaining unknowns A_1 and B_0 gives

$$A_1 = \lambda_1 \gamma \qquad B_0 = -\lambda_0 \gamma \tag{2.21}$$

where the dimensionless head difference γ is

$$\gamma = \frac{h_0^* - h_1^*}{\lambda_0 + \lambda_1} \tag{2.22}$$

The solution for the head now becomes

$$
\begin{aligned}
h &= h_0^* - \lambda_0 \gamma\, e^{x/\lambda_0} \qquad x \le 0 \\
h &= h_1^* + \lambda_1 \gamma\, e^{-x/\lambda_1} \qquad x \ge 0
\end{aligned}
\tag{2.23}
$$

Note that for the case of equal resistance values (so that $\lambda_0 = \lambda_1$), the head in the aquifer at $x = 0$ is $(h_0^* + h_1^*)/2$ as is to be expected from symmetry when the aquifer properties are the same on both sides of the dike.

The discharge vector is obtained as

$$Q_x = -T\frac{dh}{dx} \tag{2.24}$$

which gives

$$
\begin{aligned}
Q_x &= T\gamma\, e^{x/\lambda_0} \qquad x \le 0 \\
Q_x &= T\gamma\, e^{-x/\lambda_1} \qquad x \ge 0
\end{aligned}
\tag{2.25}
$$

The total flow Q_0 from the lake to the drained area is

$$Q_0 = Q_x|_{x=0} = T\gamma = T\frac{h_0^* - h_1^*}{\lambda_0 + \lambda_1} \tag{2.26}$$

Note again that 95% of Q_0 has leaked from the lake into the aquifer over a distance of $3\lambda_0$ to the left of the dike, and 95% of Q_0 has discharged from the aquifer into the drained area over a distance of $3\lambda_1$ to the right of the dike. In other words, three times the leakage factor away from a disturbance (here, the dike), the head in the aquifer is approximately equal to the head above the semi-confining layer. An example is presented below for the case that $c_1 = 4c_0$ so that $\lambda_1 = 2\lambda_0$ (Figure 2.5).

```
# parameters
h0star = 1 # head above left side leaky layer, m
h1star = -1 # head above right side leaky layer, m
k = 10 # hydraulic conductivity, m/d
H = 10 # aquifer thickness, m
c0 = 100 # resistance of left side leaky layer, d
c1 = 400 # resistance of right side leaky layer, d
T = k * H # transmissivity, m2/d
lab0 = np.sqrt(c0 * T) # leakage factor left side, m
lab1 = np.sqrt(c1 * T) # leakage factor right side, m
gamma = (h0star - h1star) / (lab0 + lab1) # dimensionless head difference, -
```

```
# solution
x = np.hstack((np.linspace(-5 * lab0, 0, 51), np.linspace(0, 5 * lab1, 51)))
h = np.zeros_like(x)
h[x < 0]  = h0star - lab0 * gamma * np.exp(x[x < 0] / lab0)
h[x >= 0] = h1star + lab1 * gamma * np.exp(-x[x >= 0] / lab1)
Qx = np.zeros_like(x)
Qx[x<0]  = T * gamma * np.exp(x[x<0] / lab0)
Qx[x>=0] = T * gamma * np.exp(-x[x>=0] / lab1)
Q0 = T * gamma
```

```
# basic plot
plt.subplot(121)
plt.plot(x, h, 'C0')
plt.subplot(122)
plt.plot(x, Qx, 'C1');
```

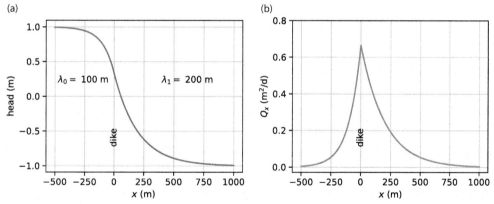

Figure 2.5 Semi-confined flow between a lake ($x \leq 0$) and a drained area ($x \geq 0$); head (left) and Q_x (right).

The stream function is contoured below. There are 20 contour intervals so that the flow between two adjacent streamlines is 5% of the flow (Figure 2.6). The first streamline that is contoured enters the aquifer at $x = -3\lambda_0$ and exits at the top of the aquifer at $x = 3\lambda_1$.

```
# solution
psi = np.zeros((2, len(x)))
psi[1] = -Qx
xg = x
zg = [0, H]
```

```
# basic streamline plot
plt.contour(xg, zg, psi, np.linspace(-Q0, 0, 21), colors='C1', linestyles='-');
```

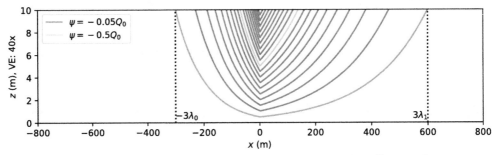

Figure 2.6   Semi-confined flow between a lake ($x \leq 0$) and a drained area ($x \geq 0$); streamlines in a vertical cross section with contour interval $\Delta \psi = 0.05 Q_0$.

Exercise 2.3: Consider one-dimensional steady flow into a long and wide lake with a leaky bottom (see Figure 2.7). A coordinate system is chosen such that $x = 0$ at the lake shore. The flow is unconfined for $x \leq 0$, while the flow is semi-confined for $x \geq 0$. Approximate the transmissivity of the aquifer as equal to T and constant everywhere. The resistance of the leaky semi-confining layer is c, and the water level in the lake is h_L. The discharge vector in the unconfined part of the aquifer is constant and is known from a water balance as $Q_x = U$. Use $h_L = 0$ m, $H = 10$ m, $T = 20$ m²/d, $U = 0.02$ m²/d, and $c = 100$ d. Derive and plot an equation for the head as a function of x. In a separate figure, draw streamlines in a vertical cross section.

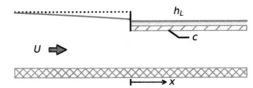

Figure 2.7   Setup for Exercise 2.3.

2.3   Flow to a long river of finite width

The next example considers one-dimensional steady flow to a long and straight river with width $2L$. A vertical cross section normal to the river is shown in Figure 2.8. The origin of the coordinate system is chosen in the middle of the river. The flow is confined to the left and right of the river and semi-confined below the river. The resistance of the leaky river bed is c, and the water level in the river is h^*. The transmissivity of the aquifer is T. The flow is uniform and equal to $Q_x = U_L$ in the aquifer to the left of the river, and the flow is uniform and equal to $Q_x = U_R$ in the aquifer to the right of the river.

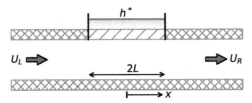

Figure 2.8   Flow to a long river of finite width.

The head and flow below the river are solved first. The boundary conditions are

$$Q_x|_{x=-L} = U_L \qquad Q_x|_{x=L} = U_R \tag{2.27}$$

The hyperbolic form of the solution (Eq. 2.9) is used for the head solution below the river

$$h - h^* = A\sinh(x/\lambda) + B\cosh(x/\lambda) \tag{2.28}$$

with the corresponding solution for the discharge vector (Eq. 2.10)

$$Q_x = -\frac{AT}{\lambda}\cosh(x/\lambda) - \frac{BT}{\lambda}\sinh(x/\lambda) \tag{2.29}$$

Substitution of the boundary conditions (Eq. 2.27) in Eq. (2.29) gives

$$A = \frac{-\lambda(U_L + U_R)}{2T\cosh(L/\lambda)} \qquad B = \frac{\lambda(U_L - U_R)}{2T\sinh(L/\lambda)} \tag{2.30}$$

The solutions for the head to the left and to the right of the river are simple linear functions, because the aquifer is confined. The solution below the river is used to compute the head $h_0 = h|_{x=-L}$ in the aquifer at the left river bank and the head $h_1 = h|_{x=L}$ in the aquifer at the right river bank. Once h_0 and h_1 are known, the solutions in the confined parts of the aquifer to the left and right of the river can be written as

$$h = -\frac{U_L}{T}(x + L) + h_0 \qquad x \le -L \tag{2.31}$$

$$h = -\frac{U_R}{T}(x - L) + h_1 \qquad x \ge L \tag{2.32}$$

An example is shown for the case that the width of the river is twice the leakage factor and the inflow at the left river bank is twice the inflow at the right river bank (Figure 2.9).

```
# parameters
k = 5 # hydraulic conductivity, m/d
H = 10 # aquifer thickness, m
hstar = 10 # head in river
L = 100 # half width of river
c = 200 # resistance of leaky river bed, d
UL = 0.1 # Qx in left aquifer
UR = -0.05 # Qx in right aquifer
T = k * H # transmissivity, m^2/d
lab = np.sqrt(T * c) # leakage factor of aquifer below river
print(f'leakage factor: {lab:.0f} m')
```

```
leakage factor: 100 m
```

```
x = np.linspace(-300, 300, 100)
A = -lab * (UL + UR) / (2 * T) / np.cosh(L / lab)
B =  lab * (UL - UR) / (2 * T) / np.sinh(L / lab)
h0 = A * np.sinh(-L / lab) + B * np.cosh(-L / lab) + hstar
h1 = A * np.sinh(L / lab) + B * np.cosh(L / lab) + hstar
h = np.zeros(len(x))
h[x <= -L] = -UL / T * (x[x <= -L] + L) + h0
h[x >= L]  = -UR / T * (x[x >= L] - L) + h1
h[np.abs(x) < L] = A * np.sinh(x[np.abs(x) < L] / lab) + \
                   B * np.cosh(x[np.abs(x) < L] / lab) + hstar
```

```
Qx = np.zeros(len(x))
Qx[x <= 0] = UL
Qx[x >= L] = UR
Qx[np.abs(x) < L] = -A * T / lab * np.cosh(x[np.abs(x) < L] / lab) - \
                     B * T / lab * np.sinh(x[np.abs(x) < L] / lab)
```

```
# basic plot
plt.subplot(121)
plt.plot(x, h, 'C0')
plt.subplot(122)
plt.plot(x, Qx, 'C1');
```

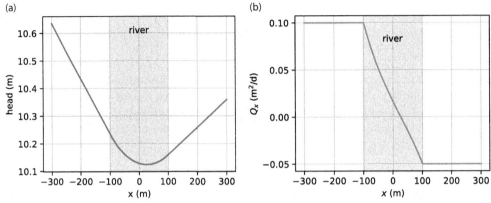

Figure 2.9 Groundwater flow to a river of finite width; head (left) and Q_x (right).

Contours of the stream function are plotted below (Figure 2.10).

```
# solution
psi = np.zeros((2, len(x)))
psi[1] = -Qx
xg = x
zg = [0, H]
```

```
# basic streamline plot
plt.contour(xg, zg, psi, np.arange(-0.1, 0.05, 0.01));
```

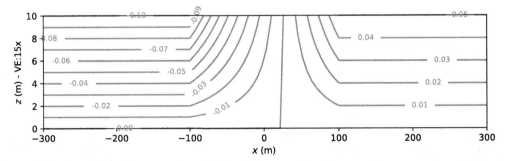

Figure 2.10 Groundwater flow to a river of finite width; streamlines in a vertical cross section with $\Delta \psi = 0.01 \, \mathrm{m}^2/\mathrm{d}$.

Exercise 2.4: Consider the case that U_L and U_R are not known, but the head at $x = -300$ m is measured to be 10.4 m and the head at $x = 300$ m is measured to be 10.6 m. Solve for U_L and U_R using a root-finding routine such as `scipy.optimize.fsolve`. Plot the head as a function of x.

2.4   Flow to a river in a two-aquifer system

Many aquifers are bounded at the bottom by a leaky layer with another aquifer below. Solutions for one-dimensional steady flow in such a two-aquifer system can be written as the sum of the solution for confined flow (Chapter 1) and the solution for semi-confined flow, as will be shown in the following.

In what follows, the transmissivity of the top aquifer (aquifer 0) is T_0 and the transmissivity of the bottom aquifer (aquifer 1) is T_1. The resistance of the leaky layer between the two aquifers is c. The recharge from rainfall to the top aquifer is equal to N. The vertical flow q_z through the leaky layer from aquifer 1 (the bottom aquifer) to aquifer 0 (the top aquifer) is

$$q_z = \frac{h_1 - h_0}{c} \tag{2.33}$$

The head in the two-aquifer system is governed by a system of two differential equations

$$\begin{aligned}
\frac{\mathrm{d}^2 h_0}{\mathrm{d}x^2} &= \frac{h_0 - h_1}{cT_0} - \frac{N}{T_0} \\
\frac{\mathrm{d}^2 h_1}{\mathrm{d}x^2} &= \frac{h_1 - h_0}{cT_1}
\end{aligned} \tag{2.34}$$

where h_0 and h_1 are the heads in the top and bottom aquifers, respectively.

The solution to the system of two differential equations (2.34) is obtained as follows. Addition of the first differential equation multiplied by T_0 and the second differential equation multiplied by T_1 gives

$$\frac{\mathrm{d}^2 (T_0 h_0 + T_1 h_1)}{\mathrm{d}x^2} = -N \tag{2.35}$$

This is Poisson's differential equation (1.8), which has the general solution

$$T_0 h_0 + T_1 h_1 = -Nx^2 + A'x + B' \tag{2.36}$$

where A' and B' are arbitrary constants. Subtraction of the second differential equation from the first differential equation gives, after a bit of algebra,

$$\frac{\mathrm{d}^2 (h_0 - h_1)}{\mathrm{d}x^2} = \frac{h_0 - h_1}{\lambda^2} - \frac{N}{T_0} \tag{2.37}$$

where

$$\lambda = \sqrt{\frac{cT_0 T_1}{T_0 + T_1}} \tag{2.38}$$

This the modified Helmholtz equation (2.5) with an additional constant $(-N/T_0)$ on the right-hand side. The general solution is

$$h_0 - h_1 = C' e^{-x/\lambda} + D' e^{x/\lambda} + \frac{N\lambda^2}{T_0} \tag{2.39}$$

where C' and D' are arbitrary constants. Expressions for h_0 and h_1 are obtained by combining Eqs. (2.36) and (2.39), which gives

$$\begin{aligned}
h_0 &= -\frac{N}{2T}x^2 + Ax + B + CT_1 e^{-x/\lambda} + DT_1 e^{x/\lambda} + \frac{T_1 N\lambda^2}{T_0 T} \\
h_1 &= -\frac{N}{2T}x^2 + Ax + B - CT_0 e^{-x/\lambda} - DT_0 e^{x/\lambda} - \frac{N\lambda^2}{T}
\end{aligned} \tag{2.40}$$

where A, B, C, and D are parameters that must be determined from boundary conditions.

As a first application, consider flow to a river in a two-aquifer system; areal recharge is neglected. The river is narrow and in direct contact with the top aquifer, but not the bottom aquifer, as shown in Figure 2.11. The boundary condition at the river is

$$h_0|_{x=0} = h_r \tag{2.41}$$

where h_r is the river stage. Far away to the left of the river, the heads in the two aquifers are equal and the head gradient is $dh_0/dx = dh_1/dx = G_L$ so that the discharge vector Q_{x_0} in the top aquifer is

$$Q_{x_0}|_{x=-\infty} = U_{L_0} = -T_0 G_L \tag{2.42}$$

and the discharge vector Q_{x_1} in the bottom aquifer is

$$Q_{x_1}|_{x=-\infty} = U_{L_1} = -T_1 G_L \tag{2.43}$$

The head gradient far to the right is G_R so that the discharge vector is

$$Q_{x_0}|_{x=\infty} = U_{R_0} = -T_0 G_R \tag{2.44}$$

and the discharge vector Q_{x_1} in the bottom aquifer is

$$Q_{x_1}|_{x=\infty} = U_{R_1} = -T_1 G_R \tag{2.45}$$

An amount σ [L^2/T] flows from the aquifer into the river per meter river normal to the plane of flow. The discharge σ is obtained from continuity of flow

$$\sigma = U_{L_0} + U_{L_1} - (U_{R_0} + U_{R_1}) \tag{2.46}$$

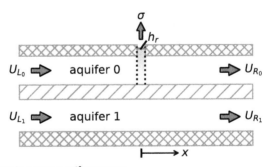

Figure 2.11   Flow to a river in a two-aquifer system.

Application of the stated boundary conditions to the general solution (2.40) gives

$$
\begin{aligned}
h_0 &= -\frac{U_{L_0}}{T_0}x - \frac{T_1 \lambda}{T_0 T}\frac{\sigma}{2}\left(e^{x/\lambda} - 1\right) + h_r \\
h_1 &= -\frac{U_{L_1}}{T_1}x + \frac{\lambda}{T}\frac{\sigma}{2}\left(e^{x/\lambda} + \frac{T_1}{T_0}\right) + h_r
\end{aligned}
\qquad x \le 0
\tag{2.47}
$$

$$h_0 = -\frac{U_{R_0}}{T_0}x - \frac{T_1\lambda}{T_0 T}\frac{\sigma}{2}\left(e^{-x/\lambda} - 1\right) + h_r$$
$$h_1 = -\frac{U_{R_1}}{T_1}x + \frac{\lambda}{T}\frac{\sigma}{2}\left(e^{-x/\lambda} + \frac{T_1}{T_0}\right) + h_r$$

$$x \geq 0 \qquad (2.48)$$

The corresponding expressions for the discharge vector are

$$Q_{x0} = U_{L_0} + \frac{T_1}{T}\frac{\sigma}{2}e^{x/\lambda}$$
$$Q_{x1} = U_{L_1} - \frac{T_1}{T}\frac{\sigma}{2}e^{x/\lambda}$$

$$x \leq 0 \qquad (2.49)$$

$$Q_{x0} = U_{R_0} - \frac{T_1}{T}\frac{\sigma}{2}e^{-x/\lambda}$$
$$Q_{x1} = U_{R_1} + \frac{T_1}{T}\frac{\sigma}{2}e^{-x/\lambda}$$

$$x \geq 0 \qquad (2.50)$$

The leakage factor λ (Eq. 2.38) for a two-aquifer system plays the same role as for semi-confined flow: A distance 3λ from the river, the heads and gradients in the aquifers are approximately equal.

In the example below, the transmissivity (and hence the flow) in the bottom aquifer is twice the transmissivity of the top aquifer. The flow is toward the river from both sides, but the flow from the left is four times as large as that from the right (Figure 2.12).

```
# parameters
k0 = 10 # hydraulic conductivity aquifer 0, m/d
k1 = 20 # hydraulic conductivity aquifer 1, m/d
H0 = 10 # thickness aquifer 0, m
H1 = 10 # thickness aquifer 1, m
Hstar = 2 # thickness leaky layer
kstar = 0.01 # hydraulic conductivity leaky layer, m/d
hr = 5 # head in river, m
GL = -0.002 # head gradient far away to the left, -
GR = 0.0005 # head gradient far away to the right, -
```

```
# solution
T0 = k0 * H0
T1 = k1 * H1
c = Hstar / kstar
T = T0 + T1
UL0 = -T0 * GL
UL1 = -T1 * GL
UR0 = -T0 * GR
UR1 = -T1 * GR
σ = UL0 + UL1 - (UR0 + UR1)
λ = np.sqrt(c * T0 * T1 / T)
print(f'leakage factor: {λ:.2f} m')
#
x = np.linspace(-4 * λ, 4 * λ, 401)
def hmaq(x):
    if x <= 0:
        h0 = -UL0 / T0 * x - T1 / T0 * λ / T * σ / 2 * (np.exp(x / λ) - 1) + hr
        h1 = -UL1 / T1 * x + λ / T * σ / 2 * (np.exp(x / λ) + T1 / T0) + hr
    else:
        h0 = -UR0 / T0 * x - T1 / T0 * λ / T * σ / 2 * (np.exp(-x / λ) - 1) + hr
```

```
        h1 = -UR1 / T1 * x + λ / T * σ / 2 * (np.exp(-x / λ) + T1 / T0) + hr
    return h0, h1

hmaqvec = np.vectorize(hmaq) # vectorized version of hmaq function
h0, h1 = hmaqvec(x)

def disxmaq(x):
    if x <= 0:
        Qx0 = UL0 + T1 / T * σ / 2 * np.exp(x / λ)
        Qx1 = UL1 - T1 / T * σ / 2 * np.exp(x / λ)
    else:
        Qx0 = UR0 - T1 / T * σ / 2 * np.exp(-x / λ)
        Qx1 = UR1 + T1 / T * σ / 2 * np.exp(-x / λ)
    return Qx0, Qx1

disxmaqvec = np.vectorize(disxmaq) # vectorized version of disxmaq function
Qx0, Qx1 = disxmaqvec(x)
```

leakage factor: 115.47 m

```
# basic plot
plt.subplot(121)
plt.plot(x, h0, 'C0', label='$h_0$')
plt.plot(x, h1, 'C0--', label='$h_1$')
plt.legend()
plt.subplot(122)
plt.plot(x, Qx0, 'C1', label='$Q_{x_0}$')
plt.plot(x, Qx1, 'C1--', label='$Q_{x_1}$')
plt.legend();
```

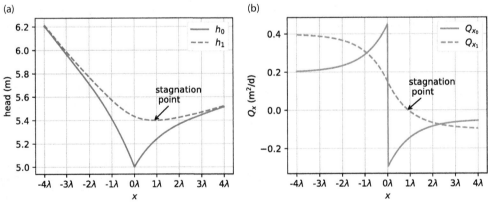

Figure 2.12 Flow to a river in a two-aquifer system; head (left) and Q_x (right).

A contour plot of the stream function is shown in Figure 2.13. Note that part of the water coming from the left in the bottom aquifer flows past the river before it leaks to the top aquifer and flows to the river. There is a stagnation point in the bottom aquifer ($Q_{x_1} = 0$) to the right of the river as can also be seen from the right graph of Figure 2.12.

```
# solution
z = [0, H1, H1 + Hstar, H1 + Hstar + H0]
psi = np.zeros((len(z), len(x)))
```

```
psi[1] = Qx1
psi[2] = Qx1
psi[3] = Qx1 + Qx0
```

```
# basic streamline plot
plt.subplot(111, aspect=10)
plt.contour(x, z, psi, np.arange(-1, 1, 0.05))
plt.axhspan(H1, H1 + Hstar, color=[.8, .8, .8]);
```

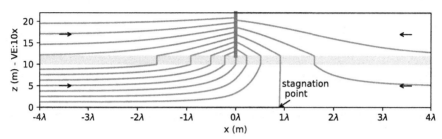

Figure 2.13 Streamlines in a vertical cross section for flow to a river in the top aquifer of a two-aquifer system.

The stated problem is fully defined by only three values: the gradients G_L and G_R far away to the left and right of the river, respectively, and the head in the river h_r. G_L may be obtained from the head in one of the aquifers on the left of the river, and G_R may be obtained from the head in one of the aquifers on the right side of the river. For example, if the head is h_d in the top aquifer at $x = -d$, substitution in Eq. (2.47) gives

$$G_L = \frac{h_r - h_d}{d} - \frac{T_1 \lambda}{T_0 T} \frac{\sigma}{2} \frac{\left(e^{-d/\lambda} - 1\right)}{d} \tag{2.51}$$

2.5   Areal recharge between two rivers in a two-aquifer system

The solution for flow between two rivers with areal recharge N in a single aquifer (Section 1.2) is expanded to two aquifers, where the rivers cut through the top aquifer, but not the bottom aquifer (see Figure 2.14). The two-aquifer solution is used to investigate when the leaky layer and the aquifer below it have a significant effect on the head in the top aquifer or when the bottom of the top aquifer may be approximated as impermeable. The origin is chosen halfway between the two rivers, and the distance between the two rivers is $2L$. The heads in the two rivers are equal to h_L so that the boundary conditions in the top aquifer are

$$h_0|_{x=-L} = h_0|_{x=L} = h_L \tag{2.52}$$

The horizontal flow below the rivers in the bottom aquifer is approximated as zero, which is the case when discharge into the rivers is the same from both sides of the river

$$\frac{\mathrm{d}h_1}{\mathrm{d}x}\bigg|_{x=-L} = \frac{\mathrm{d}h_1}{\mathrm{d}x}\bigg|_{x=L} = 0 \tag{2.53}$$

Application of the boundary conditions (2.52) and (2.53) to the general solution for flow in a two-aquifer system (2.40), but using hyperbolic sine and cosine functions rather than

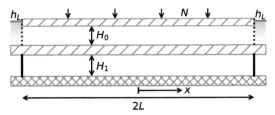

Figure 2.14 Infiltration between two rivers in a two-aquifer system.

exponential functions, results in the following solution for the head

$$h_0 = -\frac{N}{2T}(x^2 - L^2) - \frac{NL\lambda T_1}{TT_0}\frac{[\cosh(x/\lambda) - \cosh(L/\lambda)]}{\sinh(L/\lambda)} + h_L \qquad (2.54)$$

$$h_1 = -\frac{N}{2T}(x^2 - L^2) + \frac{NL\lambda}{TT_0}\frac{[T_0\cosh(x/\lambda) - T_1\cosh(L/\lambda)]}{\sinh(L/\lambda)} - \frac{N\lambda^2}{T_0} + h_L \qquad (2.55)$$

The corresponding discharge vector in the two aquifers is

$$Q_{x_0} = -T_0\frac{dh_0}{dx} = \frac{T_0 Nx}{T} + \frac{NLT_1\sinh(x/\lambda)}{T\sinh(L/\lambda)} \qquad (2.56)$$

$$Q_{x_1} = -T_1\frac{dh_1}{dx} = \frac{T_1 Nx}{T} - \frac{NLT_1\sinh(x/\lambda)}{T\sinh(L/\lambda)} \qquad (2.57)$$

As an example, the head and flow in the two-aquifer system are simulated for the case that the leakage factor λ is equal to 20% of the distance between the two rivers ($2L = 5\lambda$). The head is also simulated with a single-aquifer model, using the solution of Section 1.2. The head in the top aquifer of the two-aquifer model is smaller than the head in the single-aquifer model (Figure 2.15). The relative difference is at most 27% for the values used below. Water that infiltrates in the center part of the system flows partly through the bottom aquifer on its way to the river.

```
# parameters
hL = 0 # head in river, m
N = 0.001 # recharge, m/d
L = 500 # half distance between two rivers, m
k0 = 10 # hydraulic conductivity aquifer 0, m/d
H0 = 10 # thickness aquifer 0, m
k1 = 20 # hydraulic conductivity aquifer 1, m/d
H1 = 10 # thickness aquifer 1, m
kv = 0.005 # hydraulic conductivity leaky layer, m/d
Hv = 3 # thickness leaky layer, m
T0 = k0 * H0 # transmissivity aquifer 0, m^2/d
T1 = k1 * H1 # transmissivity aquifer 1, m^2/d
T = T0 + T1
c = Hv / kv # resistance leaky layer, d
lab = np.sqrt(c * T0 * T1 / (T0 + T1)) # leakage factor
print(f'leakage factor: {lab:.0f} m')
```

```
leakage factor: 200 m
```

```
# solution head
x = np.linspace(-L, L, 401)
h0 = -N / (2 * T) * (x ** 2 - L ** 2) - N * L * lab * T1 / (T * T0) * (
    np.cosh(x / lab) - np.cosh(L / lab)) / np.sinh(L / lab) + hL
```

```
h1 = -N / (2 * T) * (x ** 2 - L ** 2) + N * L * lab  / (T * T0) * (
    T0 * np.cosh(x / lab) + T1 * np.cosh(L / lab)) / np.sinh(L / lab) -\
    N * lab ** 2 / T0 + hL
h = -N / (2 * T0) * (x ** 2 - L ** 2) + hL # single-layer solution, Section 1.2
print('relativediff in head at center between single-layer and two-layer sol:',
    f'{(h[200] - h0[200]) / h0[200]:.2f}')
# solution Qx
xg = np.linspace(-L, L, 400)
Qx0 = T0 * N * xg / T + N * L * T1 * np.sinh(xg / lab) / (T * np.sinh(L / lab))
Qx1 = T1 * N * xg / T - N * L * T1 * np.sinh(xg / lab) / (T * np.sinh(L / lab))
Qx = N * xg
# solution stream function
zg = [-H0 - Hv - H1, -H0 - Hv, -H0, 0]
psi = np.zeros((4, len(xg)))
psi[1] = Qx1
psi[2] = Qx1
psi[3] = Qx1 + Qx0
```

relative diff in head at center between single-layer and two-layer sol: 0.27

```
# basic plot
plt.subplot(121)
plt.plot(x, h0, label='aquifer 0')
plt.plot(x, h1, label='aquifer 1')
plt.plot(x, h, 'k--', label='single aquifer')
plt.legend()
plt.subplot(122)
plt.plot(xg, Qx0, label='aquifer 0')
plt.plot(xg, Qx1, label='aquifer 1')
plt.plot(xg, Qx, 'k--', label='single aquifer');
```

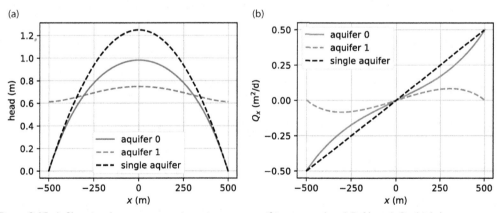

Figure 2.15 Infiltration between two rivers in a two-aquifer system; head (left) and Q_x (right).

Streamlines in a vertical cross section are shown in Figure 2.16.

```
# basic streamline plot
plt.contour(xg, zg, psi, 40);
```

The difference between a single-layer model with an impermeable base and a two-aquifer model increases when the leakage factor λ decreases compared to the distance L between the two rivers. The leakage factor increases when, for example, the resistance c of the leaky layer

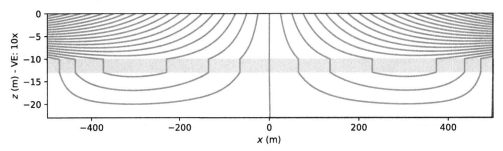

Figure 2.16 Streamlines in a vertical cross section for infiltration between two rivers in a two-aquifer system.

is larger or when the transmissivity of the bottom aquifer is larger, as can be seen from Eq. 2.38. The head halfway the two rivers ($x = 0$) in the top aquifer is plotted for different values of the leakage factor (Figure 2.17). The dashed line represents the head at the center of the aquifer when the bottom of the top aquifer is approximated as impermeable.

```
lablist = np.linspace(100, 1000, 100)
hcenter = np.zeros(len(lablist))
for i, lab in enumerate(lablist):
    hcenter[i] = N / (2 * T) * L ** 2 - N * L * lab * T1 / (T * T0) * (
                 1 - np.cosh(L / lab)) / np.sinh(L / lab) + hL
```

```
# basic plot
plt.plot(lablist, hcenter)
plt.axhline(N * L ** 2 / (2 * T0), color='k', ls='--');
```

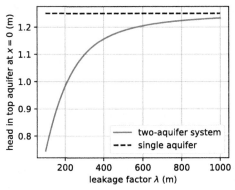

Figure 2.17 Head halfway the two rivers ($x = 0$) in the top aquifer as a function of the leakage factor, and the head for a single aquifer with an impermeable base.

Exercise 2.5: Compute the effective transmissivity of the single-aquifer solution such that the head at $x = 0$ is the same as the head at $x = 0$ in the two-aquifer solution. Plot in one graph the head vs. x in the single aquifer using the effective transmissivity and the head vs. x in the top aquifer of the two-aquifer solution.

Chapter 3

Steady one-dimensional unconfined flow with variable saturated thickness

In the previous chapters, the saturated thickness of the aquifer was approximated as constant. The approximation of a constant saturated thickness gives reasonable results when the variation of the saturated thickness is not too large. In this chapter, solutions are derived for unconfined flow with a saturated thickness that varies with the head. The Dupuit–Forchheimer approximation is adopted so that the head is equal to the elevation of the water table and the horizontal flow is distributed uniformly over the saturated thickness (Figure 3.1).

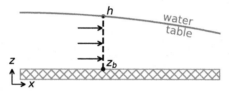

Figure 3.1 Unconfined Dupuit–Forchheimer flow over a horizontal base.

The elevation of the base of the aquifer is denoted as z_b so that the saturated thickness is equal to $h - z_b$ (Figure 3.1) and the discharge vector can be written as

$$Q_x = -k(h - z_b)\frac{dh}{dx} = -\frac{d}{dx}\left(\tfrac{1}{2}k(h - z_b)^2\right) \tag{3.1}$$

where it is used that the elevation of the base of the aquifer and the hydraulic conductivity are constant.

While it is possible to obtain a solution for unconfined flow in terms of the head h, it is more convenient to solve unconfined flow problems using a discharge potential. The discharge potential for unconfined flow Φ is introduced as

$$\Phi = \tfrac{1}{2}k(h - z_b)^2 \tag{3.2}$$

so that the discharge vector can be written as

$$Q_x = -\frac{d\Phi}{dx} \tag{3.3}$$

Flow fields where the discharge vector Q_x may be written as minus the derivative of a discharge potential are referred to as potential flow. Potential flow solutions exist in many fields, including fluid mechanics and heat transport.

Discharge potentials are an abstract quantity that cannot be measured in the field. Groundwater problems are, of course, not formulated in terms of discharge potentials, but in terms of heads. To be able to use a potential solution, boundary conditions in terms of heads must be

DOI: 10.1201/9781315206134-3

converted to boundary conditions in terms of discharge potentials. After a potential solution is evaluated, the computed potential must be converted back to a head. The use of discharge potentials makes it possible to solve fairly complicated problems, including unconfined flow and interface flow in coastal aquifers (Chapter 4).

The continuity equation for one-dimensional flow is still Eq. (1.5)

$$\frac{dQ_x}{dx} = N \tag{3.4}$$

Substitution of Eq. (3.3) for Q_x into the continuity equation gives

$$\frac{d^2\Phi}{dx^2} = -N \tag{3.5}$$

The general solution to this second-order ordinary differential equation may be written as

$$\Phi = -\frac{N}{2}x^2 + Ax + B \tag{3.6}$$

where the constants A and B need to be determined from the boundary conditions. Once a solution for the discharge potential is found, the solution for the head is obtained with the inverse of Eq. (3.2)

$$h = z_b + \sqrt{2\Phi/k} \tag{3.7}$$

3.1   Areal recharge between an impermeable boundary and a river

Consider uniform recharge at a rate N on an unconfined aquifer bounded on the left by an impermeable boundary and on the right by a river (see Figure 3.2). The problem is the same as in Section 1.3, but now the transmissivity is a function of the head h in the aquifer rather than approximated as constant. The length of the aquifer is L, and the elevation of the aquifer bottom is z_b. The boundary conditions are

$$\frac{dh}{dx}\bigg|_{x=0} = 0 \tag{3.8}$$

$$h|_{x=L} = h_L \tag{3.9}$$

where h_L is the head in the river at $x = L$.

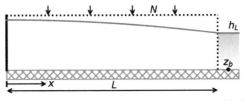

Figure 3.2   Areal recharge on an unconfined aquifer between an impermeable boundary and a river.

The boundary conditions in terms of heads (Eqs. 3.8 and 3.9) are converted to boundary conditions in terms of the discharge potential for unconfined flow

$$\frac{d\Phi}{dx}\bigg|_{x=0} = 0 \tag{3.10}$$

$$\Phi|_{x=L} = \Phi_L = \tfrac{1}{2}k(h_L - z_b)^2 \tag{3.11}$$

Application of the boundary conditions to the general solution (3.6) gives

$$\Phi = -\frac{N}{2}(x^2 - L^2) + \Phi_L \tag{3.12}$$

The head is obtained from the potential with Eq. (3.7), which gives, using Eq. (3.11)

$$h = z_b + \sqrt{2\Phi/k} = z_b + \sqrt{-N(x^2 - L^2)/k + (h_L - z_b)^2} \tag{3.13}$$

The discharge vector Q_x is obtained by differentiation of the potential as

$$Q_x = -\frac{d\Phi}{dx} = Nx \tag{3.14}$$

An example is presented below. The solution is compared to solution (1.39), where the saturated thickness was approximated as $H = 10$ m (Figure 3.3). The specific choice of the approximate H in the example leads to an underestimate of the head in the aquifer at the impermeable boundary. The discharge vector, on the other hand, is the same for both problems (compare Eqs. (3.14) and (1.40)), because Q_x is independent of the choice of the approximate thickness H.

```
# parameters
L = 1000 # aquifer length, m
H = 10 # approximate aquifer thickness used in solution of Section 1.3, m
zb = -5 # aquifer bottom, m
k = 10 # hydraulic conductivity, m/d
n = 0.3 # porosity, -
hL = 4 # specified head at the right boundary, m
N = 0.001  # areal recharge, m/d
```

```
# solution
phiL = 0.5 * k * (hL - zb) ** 2
x = np.linspace(0, L, 100)
phi = -N / 2 * (x ** 2 - L ** 2) + phiL
h = zb + np.sqrt(2 * phi / k)
happrox = -N / (2 * k * H) * (x ** 2 - L ** 2) + hL
Qx = N * x
```

```
# basic plot
plt.subplot(121)
plt.plot(x, h, 'C0')
plt.plot(x, happrox, 'C0--')
plt.subplot(122)
plt.plot(x, Qx, 'C1');
```

```
print(f'flux to left river: {-Qx[0]:.3f} m^2/d')
print(f'flux to right river: {Qx[-1]:.3f} m^2/d')
```

```
flux to left river: -0.000 m^2/d
flux to right river: 1.000 m^2/d
```

A plot of the stream function may also be created, but now the top of the aquifer is formed by the water table (Figure 3.4).

```
# solution
psi = np.zeros((2, len(x)))
```

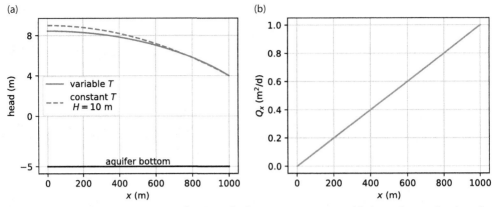

Figure 3.3 Areal recharge on an unconfined aquifer between an impermeable boundary and a river; head (left) and Q_x (right).

```
psi[1] = -Qx
xg = np.zeros_like(psi)
xg[:] = x
zg = np.zeros_like(psi)
zg[0] = zb
zg[1] = h
```

```
# basic streamline plot
plt.subplot(111, aspect=25)
plt.contour(xg, zg, psi, 10, colors='C1', linestyles='-')
plt.plot(x, h, 'k');
```

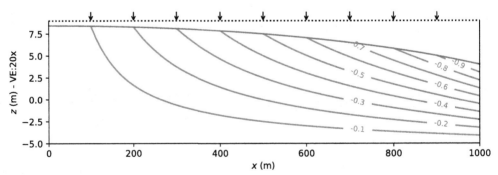

Figure 3.4 Streamlines in a vertical cross section for areal recharge on an unconfined aquifer between two rivers.

The travel time is computed for water that reaches the water table at position x and flows to the river at $x = L$. The horizontal component of the velocity vector is obtained as

$$v_x = \frac{Q_x}{n(h - z_b)} = \frac{\sqrt{Nk}}{n} \frac{x}{\sqrt{2\Phi_L/N + L^2 - x^2}} \tag{3.15}$$

The travel time can be computed, as before, as

$$t_{\text{tr}}(x, L) = \int_x^L \frac{\mathrm{d}x}{v_x} \tag{3.16}$$

The integration is conducted analytically and gives (omitting the constant of integration)

$$F(x) = \int \frac{\mathrm{d}x}{v_x} = \frac{n}{\sqrt{Nk}} \sqrt{a - x^2} - \sqrt{a}\, \mathrm{arctanh}\left(\frac{\sqrt{a - x^2}}{\sqrt{a}}\right) \qquad (3.17)$$

where

$$a = 2\Phi_L/N + L^2 \qquad (3.18)$$

The travel time can now be computed as

$$t_{\mathrm{tr}}(x, L) = F(L) - F(x) \qquad (3.19)$$

The travel time is plotted from point x to the river in Figure 3.5.

```
# solution
def integral(x):
    a = 2 * phiL / N + L ** 2
    return np.sqrt(a - x ** 2) - np.sqrt(a) * \
    np.arctanh(np.sqrt(a - x ** 2) / np.sqrt(a))

def traveltime(x):
    return n / np.sqrt(N * k) * (integral(L) - integral(x))

x = np.linspace(10, L, 100)
trtime = traveltime(x)
```

```
# basic travel time plot
plt.plot(x, trtime);
```

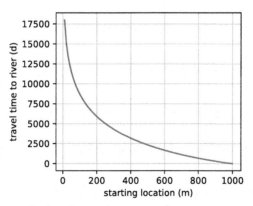

Figure 3.5 Travel time from starting location to river at $x = L$.

Exercise 3.1: Determine the value of the effective saturated thickness H such that the head at the impermeable boundary is the same for the solutions with variable saturated thickness and constant saturated thickness. Plot the head vs. x for both solutions.

Exercise 3.2: Compare the travel time from $x = 400$ to the river computed with the analytic solution derived above and numerical integration of the integral (Eq. 3.16) using the quad function of scipy.integrate.

Exercise 3.3: Compute the travel time for water that recharges the aquifer at position x to reach the river at $x = L$ for the case that the saturated thickness is approximated as constant and equal to $H = 10$ m, and add the computed travel time to Figure 3.5.

Exercise 3.4: Consider the case of unconfined flow between two parallel rivers as in Section 1.2, but now the saturated thickness of the unconfined aquifer is not approximated by an effective thickness H, but varies with the head instead. Compute the difference between the solution for unconfined flow with a varying saturated thickness and the solution for unconfined flow with an effective thickness H. Use the parameters specified in Section 1.2. Compare the location of the water divide and the maximum head between the rivers for the two solutions.

3.2 Flow over a step in the aquifer base

The problem of unconfined flow between two parallel rivers without areal recharge is considered, but this time, there is a step in the base of the aquifer as shown in Figure 3.6. The base of the aquifer is equal to z_0 for the left part of the aquifer (length L_0) and is equal to z_1 for the right part of the aquifer (length L_1). The head is h_0 at $x = 0$, and the head is h_L at $x = L$. These conditions are transformed to conditions in terms of the discharge potential

$$\Phi|_{x=0} = \Phi_0 = \tfrac{1}{2}k(h_0 - z_0)^2 \tag{3.20}$$

$$\Phi|_{x=L} = \Phi_L = \tfrac{1}{2}k(h_L - z_1)^2 \tag{3.21}$$

As for the solution of, e.g., Section 1.4, the head and flux are continuous at $x = L_0$.

$$h^- = h^+ \tag{3.22}$$

$$Q_x^- = Q_x^+ \tag{3.23}$$

where the superscripts $^-$ and $^+$ are used as shorthand for evaluation just to the left and right of $x = L_0$, respectively.

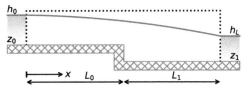

Figure 3.6 Unconfined flow with a step in the base.

The flow from the left river to the right river is equal to U, which is unknown upfront. Application of boundary conditions (3.20) and (3.21) gives equations for the discharge potential for the left and right parts that still contain the unknown flux U

$$\Phi = -Ux + \Phi_0 \qquad 0 \le x \le L_0 \tag{3.24}$$

$$\Phi = -U(x - L) + \Phi_L \qquad L_0 \le x \le L \tag{3.25}$$

where $L = L_0 + L_1$. The discharge U is obtained from condition (3.22) as follows. The discharge potentials Φ^- and Φ^+ at $x = L_0$ are

$$\Phi^- = -UL_0 + \Phi_0 = \tfrac{1}{2}k(h^- - z_0)^2 \tag{3.26}$$

$$\Phi^+ = -U(L_0 - L) + \Phi_L = \tfrac{1}{2}k(h^+ - z_1)^2 \tag{3.27}$$

which are transformed in the following expressions for h^- and h^+.

$$h^- = \sqrt{\frac{2(-UL_0 + \Phi_0)}{k}} + z_0 \tag{3.28}$$

$$h^+ = \sqrt{\frac{2(UL_1 + \Phi_L)}{k}} + z_1 \tag{3.29}$$

Setting $h^- = h^+$ gives an equation from which U can be determined, but it is not possible to write an explicit equation for U. The value of U for which $h^- = h^+$ is determined numerically. First, h^- and h^+ are plotted vs. U in Figure 3.7.

```
# parameters
k = 10 # hydraulic conductivity, m/d
z0 = 0 # base elevation left section, m
z1 = -4 # base elevation right section, m
L0 = 500 # length of left section, m
L1 = 500 # length of right section, m
L = L0 + L1 # total distance between rivers, m
h0 = 10 # specified head at the left boundary, m
hL = 0 # specified head at the right boundary, m
```

```
# solution
phi0 = 0.5 * k * (h0 - z0)**2
phiL = 0.5 * k * (hL - z1)**2

def hmin(U, L0=L0, z0=z0, phi0=phi0):
    return np.sqrt(2 * (-U * L0 + phi0) / k) + z0

def hplus(U, L1=L1, z1=z1, phiL=phiL):
    return np.sqrt(2 * (U * L1 + phiL) / k) + z1
```

```
# basic plot two conditions
U = np.linspace(0, 1, 100)
plt.plot(U, hmin(U), label='$h^-$')
plt.plot(U, hplus(U), label='$h^+$')
plt.legend();
```

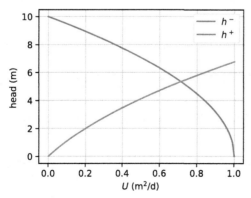

Figure 3.7 Head at $x = L_0$ vs. U.

As can be seen from the graph in Figure 3.7, the lines for h^- and h^+ intersect somewhere between $U = 0.6$ m²/d and $U = 0.8$ m²/d. To determine the intersection point, a Python function is written to compute the difference between h^- and h^+ for a given value of U. The value of U for which this function is zero (so that $h^- = h^+$) is found with the fsolve method of the scipy.optimize package.

```
from scipy.optimize import fsolve

def hdiff(U):
    return hmin(U) - hplus(U)

U = fsolve(hdiff, 0.7)[0]  # first value of array returned by fsolve
print(f'U: {U:0.4f} m^2/d')
```

U: 0.7139 m^2/d

The head can be computed with Eq. (3.28) and (3.29) now that U is known. The water table and streamlines in a vertical cross section are plotted in Figure 3.8. Note that the water table is continuous, but its slope (i.e., the head gradient) is not. This makes sense as the saturated thickness suddenly increases at the change in base while the discharge vector is continuous. This is only possible if at the same point the head gradient suddenly decreases. The elevations of the streamlines in Figure 3.8 jump at the change in the base. Both the jump in head gradient and the jumps in the streamlines are a consequence of the Dupuit–Forchheimer approximation, which means that streamlines can shift up or down without the loss of energy. In reality, the transition of both the head and the streamlines across a step in the base is smooth, as is demonstrated in Section 10.3.

```
# solution
x = np.hstack((np.linspace(0, L0 - 1e-6, 100), np.linspace(L0 + 1e-6, L, 100)))
phi = np.empty_like(x)
phi[x < L0] = -U * x[x < L0] + phi0
phi[x >= L0] = -U * (x[x >= L0] - L) + phiL
h = np.zeros_like(phi)
h[x < L0] = np.sqrt(2 * phi[x < L0] / k) + z0
h[x >= L0] = np.sqrt(2 * phi[x >= L0] / k) + z1
#
psi = np.zeros((2, len(x)))
psi[1] = -U
xg = np.zeros_like(psi)
xg[:] = x
zg = np.zeros_like(xg)
zg[0, :100] = z0
zg[0, 100:] = z1
zg[1] = h
```

```
# basic streamline plot
plt.subplot(111, aspect=25)
plt.contour(xg, zg, psi, np.linspace(-U, 0, 4), colors='C1', linestyles='-')
plt.plot(x, h, 'C0')
plt.plot(x, zg[0], 'k');
```

3.3   Combined confined/unconfined flow with areal recharge

So far in this chapter, separate solutions have been derived for confined flow (or flow with constant transmissivity) and unconfined flow. In this section, a case is considered where the flow changes from confined flow to unconfined flow somewhere in the aquifer. Consider flow between two parallel rivers in a confined aquifer with thickness H (Figure 3.9); the rivers are

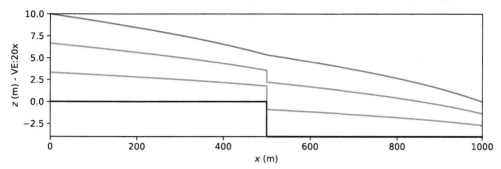

Figure 3.8 Streamlines in a vertical cross section for unconfined flow over a step in the base; $\Delta\psi = U/3$.

in direct contact with the aquifer. The rivers are a distance L apart, and the areal recharge is uniform and equal to N. The stage h_0 in the left river is above the top of the aquifer, and the stage h_L in the right river is below the top of the aquifer. Hence, groundwater flow is confined near the left river, is unconfined near the right river, and changes somewhere in the aquifer from confined to unconfined.

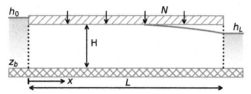

Figure 3.9 Combined confined/unconfined flow between two rivers.

One approach to solving this problem is to write one solution for the head in the confined part of the aquifer and a separate solution for the head in the unconfined part and then try to find the point where flow changes from confined to unconfined. This is possible, but the use of the discharge potential provides an easier way. The discharge potential was already defined for unconfined flow, but not yet for confined flow. For confined flow, the discharge vector may be written as

$$Q_x = -kH\frac{\mathrm{d}h}{\mathrm{d}x} = -\frac{\mathrm{d}}{\mathrm{d}x}(kHh) = -\frac{\mathrm{d}\Phi}{\mathrm{d}x} \tag{3.30}$$

where both k and H are constant. The discharge potential for confined flow is defined as

$$\Phi = kHh + C \tag{3.31}$$

where C is an arbitrary constant. When the flow is purely confined, the constant C can be set to zero, but when the flow is confined in parts of the aquifer and unconfined in other parts of the aquifer, the constant C must be picked differently, as explained in the following.

Now that both confined flow and unconfined flow are formulated as potential flow, the discharge vector may be written as Eq. (3.3), the governing differential equation is the Poisson equation (3.5), and the general solution for one-dimensional steady potential flow is Eq. (3.6). This means that the discharge potential is a continuous function of x, irrespective of whether the flow is confined or unconfined. The discharge potential must be converted to a head that is also a continuous function of x (as the head cannot suddenly jump up or down without a physical barrier to flow). This is achieved by proper selection of the constant C in Eq. (3.31). At the point where the flow transitions from confined to unconfined, the head in the aquifer is equal to $z_b + H$. The potential at this point is referred to as the transition potential Φ_t and may be computed from the potential for unconfined flow (Eq. 3.2) as

$$\Phi_t = \tfrac{1}{2}kH^2 \tag{3.32}$$

The constant C is chosen such that the discharge potential for confined flow yields the same transition potential

$$\Phi_t = kH(z_b + H) + C = \tfrac{1}{2}kH^2 \tag{3.33}$$

so that

$$C = -\tfrac{1}{2}kH^2 - kHz_b \tag{3.34}$$

Once the discharge potential is computed as a function of x, it is converted to a head using the appropriate equation. If the flow is confined,

$$h = \frac{\Phi - C}{kH} \qquad \Phi \geq \Phi_t \tag{3.35}$$

and if the flow is unconfined,

$$h = z_b + \sqrt{2\Phi/k} \qquad \Phi \leq \Phi_t \tag{3.36}$$

As an example, the problem of Section 1.2 is solved, but now the thickness of the aquifer is fixed to H while the saturated thickness of the aquifer may vary with the head. The boundary conditions are

$$\Phi|_{x=0} = \Phi_0 = kHh_0 + C \tag{3.37}$$
$$\Phi|_{x=L} = \Phi_L = \tfrac{1}{2}k(h_L - z_b)^2 \tag{3.38}$$

Application of the boundary conditions to the general solution (3.6) gives

$$\Phi = -\frac{N}{2}(x^2 - Lx) + \frac{(\Phi_L - \Phi_0)x}{L} + \Phi_0 \tag{3.39}$$

The head solution is obtained from the potential with Eq. (3.35) or (3.36). Results for the same parameters as Section 1.2 are shown in Figure 3.10. Note that the flow is unconfined for only a small section of the aquifer near the right river.

```
# parameters
L = 1000 # aquifer length, m
H = 10 # aquifer thickness, m
zb = -5 # aquifer base, m
k = 10 # hydraulic conductivity, m/d
h0 = 6 # specified head at the left boundary, m
hL = 4 # specified head at the right boundary, m
N = 0.001  # areal recharge, m/d
```

```
# solution
C = -0.5 * k * H**2 - k * H * zb
phi0 = k * H * h0 + C
phi1 = 0.5 * k * (hL - zb)**2
phit = 0.5 * k * H**2  # transition potential
x = np.linspace(0, L, 400)
```

```
phi = -N / 2 * (x ** 2 - L * x) + (phi1 - phi0) * x / L + phi0
h = np.zeros_like(phi)
h[phi >= phit] = (phi[phi > phit] - C) / (k * H)
h[phi <= phit] = zb + np.sqrt(2 * phi[phi <= phit] / k)
Qx = N * (x - L / 2) - (phi1 - phi0) / L
happrox = -N / (2 * k * H) * (x ** 2 - L * x) + (hL - h0) * x / L + h0
Qxapprox = N * (x - L / 2) - k * H * (hL - h0) / L
```

```
# basic plot
plt.subplot(121)
plt.plot(x, h, 'C0')
plt.plot(x, happrox, 'C0--')
plt.subplot(122)
plt.plot(x, Qx, 'C1')
plt.plot(x, Qxapprox, 'C1--')
```

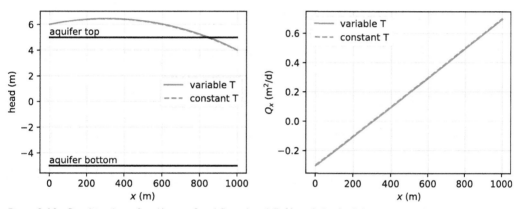

Figure 3.10 Combined confined/unconfined flow; head (left) and Q_x (right).

The flux into the left river is now slightly higher than in Section 1.2, while the flux into the right river is slightly lower, so that the sum of the outflow in the two rivers is still equal to *NL*.

A contour plot of the steam function is created below (Figure 3.11). The top of the saturated zone is formed by the top of the aquifer in the zone where the flow is confined, and by the water table in the zone where the flow is unconfined.

```
print(f'flux to left river: {-Qx[0]:.3f} m^2/d')
print(f'flux to right river: {Qx[-1]:.3f} m^2/d')
```

```
flux to left river: 0.305 m^2/d
flux to right river: 0.695 m^2/d
```

```
# solution
psi = np.zeros((2, len(x)))
psi[1] = -Qx
xg = np.zeros_like(psi)
xg[:] = x
zg = np.zeros_like(psi)
zg[0] = zb
zg[1] = H + zb
zg[1, h < H + zb] = h[h < H + zb]
```

```
# basic streamline plot
plt.subplot(111, aspect=25)
plt.contour(xg, zg, psi, 20, colors='C1', linestyles='-')
plt.plot(xg[0], zg[1], 'C0');
```

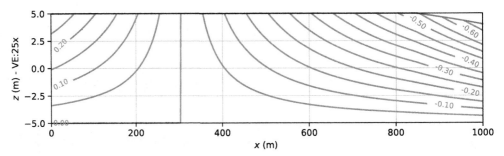

Figure 3.11 Streamlines in a vertical cross section for combined confined/unconfined flow.

Exercise 3.5: Consider the case that $h_0 = h_L = 4$ m and $N = 0.002$ m/d. All other parameters are as for the example above, except for the aquifer thickness. Compute the head halfway between the two rivers for $H = 10$ m and for $H = 20$ m. Plot the head as a function of x for both cases in the same graph.

Chapter 4

Steady one-dimensional flow in coastal aquifers

The salinity (C_s) of groundwater in coastal aquifers ranges from almost zero for freshwater to $C_s = 36$ g/L for intruded seawater. Because freshwater has a lower density ($\rho_f = 1000$ kg/m^3) than seawater ($\rho_s = 1025$ kg/m^3), fresh groundwater tends to float on top of saline groundwater. The water density ρ at $20°C$ may be computed from the salinity using the empirical formula

$$\rho = \rho_f + 0.7 C_s \tag{4.1}$$

where C_s is in g/L and ρ in kg/m^3. This approximation is reasonable for salt concentrations up to the salinity of seawater.

In coastal aquifers, the freshwater is separated from the saltwater by a brackish transition zone. When the transition zone is relatively thin, that is, much thinner than the thickness of the freshwater lens, it can be approximated by an interface. An interface is a boundary between fluids that do not mix. Replacing the transition zone (in which there is mixing) by an interface greatly simplifies the mathematical treatment of variable-density flow problems and provides a means to develop analytical solutions for coastal aquifer flow. The location where the interface is at the top of the aquifer is called the tip of the interface, while the point where the interface is at the bottom of the aquifer is called the toe of the interface (Figure 4.1).

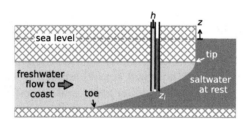

Figure 4.1 Steady interface in a coastal aquifer, where h is the head in the freshwater zone above sea level and z_i is the elevation of the interface with respect to sea level.

Steady interface flow means that the interface has reached a steady-state position, the saltwater is at rest, and only the freshwater is flowing. When the saltwater is at rest and an observation well is screened in the saltwater, the saltwater rises to exactly the sea level; this is called the saltwater head. As the saltwater is not flowing, the saltwater head in the saltwater is the same everywhere. After all, Darcy's law states that water moves from a high head to a low head. When the Dupuit approximation is adopted for flow in the freshwater zone, the depth of the interface may be computed from the head in the freshwater as explained in the following.

Consider steady interface flow as depicted in Figure 4.1. The elevation of the interface z_i is measured with respect to the sea level (i.e., z_i is negative). An observation well has its opening

DOI: 10.1201/9781315206134-4

at the interface and is filled with freshwater. The freshwater rises to elevation h above sea level so that the pressure p_f at the bottom of the observation well is

$$p_f = \rho_f g(h - z_i) \tag{4.2}$$

where ρ_f is the density of freshwater. Just to the right of the freshwater observation well is a second observation well, which also has its opening at the interface but is filled with saltwater. The saltwater in the second observation well rises up to sea level ($z = 0$) (Figure 4.1). The pressure p_s at the bottom of the second observation well is

$$p_s = \rho_s g(0 - z_i) \tag{4.3}$$

The pressures at the bottoms of both observation wells must be equal ($p_f = p_s$ since they are at the same elevation and the interface is not moving). Setting Eq. (4.2) and Eq. (4.3) equal and rearrangement of terms leads to an expression for the elevation of the interface z_i as a function of the head h above sea level in the freshwater

$$z_i = -\alpha h \tag{4.4}$$

where

$$\alpha = \frac{\rho_f}{\rho_s - \rho_f} \tag{4.5}$$

Relation (4.4) is known as the Ghyben–Herzberg equation (e.g., Post et al., 2018). Using the density values for freshwater and seawater provided above gives $\alpha = 40$. In words: If the head in the freshwater at a location is 1 m above sea level, the steady interface at that location is 40 m below sea level.

Note that the Ghyben–Herzberg relation holds for the case that the freshwater flow is steady and the saltwater is at rest. The head h in Eq. (4.4) is the head in the freshwater at the interface. When the Dupuit approximation is adopted (i.e., the resistance to flow in the vertical direction is neglected and the vertical head distribution is approximated as hydrostatic), the elevation of the interface may be computed from the head at any elevation in the freshwater with the Ghyben–Herzberg equation (4.4).

4.1 Confined interface flow

Consider steady interface flow in a confined coastal aquifer as shown in Figure 4.2. The elevations of the top and bottom of the aquifer are z_t and z_b, respectively. The thickness of the aquifer is $H = z_t - z_b$. Flow problems in coastal aquifers are commonly broken up into a part near the coast where the interface is present and a part upstream of the interface toe where there is no interface. The flow in both parts may be solved simultaneously by defining an appropriate discharge potential such that the flow may be computed in both parts of the aquifer as

$$Q_x = -\frac{d\Phi}{dx} \tag{4.6}$$

This was first done by Strack (1976), and these discharge potentials are also referred to as Strack potentials. Separate discharge potentials must be defined for the two parts of the flow system. Note that this approach is similar to the approach for combined confined/unconfined flow (Section 3.3), where separate discharge potentials were defined for the confined and the unconfined parts of the flow system.

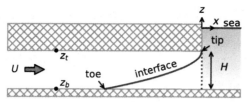

Figure 4.2 Interface flow toward the coast in a confined aquifer.

First, consider the interface flow part. The elevation of the tip of the interface is at the top of the aquifer $z_{\text{tip}} = z_t$, which corresponds to a head $h_{\text{tip}} = -z_t/\alpha$. The elevation of the toe of the interface is $z_{\text{toe}} = z_b$, which corresponds to a head $h_{\text{toe}} = -z_b/\alpha$. Between the tip and the toe, the thickness of the freshwater zone is equal to $z_t - z_i = z_t + \alpha h$. The discharge vector may be written as

$$Q_x = -k(z_t + \alpha h)\frac{\text{d}h}{\text{d}x} = -k\alpha(h + z_t/\alpha)\frac{\text{d}h}{\text{d}x} = -\frac{\text{d}\Phi}{\text{d}x} \tag{4.7}$$

so that the discharge potential is

$$\Phi = \tfrac{1}{2}k\alpha(h + z_t/\alpha)^2 \qquad -z_t/\alpha \le h \le -z_b/\alpha \tag{4.8}$$

The potential Φ_{toe} at the toe is obtained by substituting the head h_{toe} for h in Eq. (4.8)

$$\Phi_{\text{toe}} = \frac{kH^2}{2\alpha} \tag{4.9}$$

Upstream of the toe, the flow is regular confined flow so that the saturated thickness is equal to H and the discharge vector may be written as

$$Q_x = -kH\frac{\text{d}h}{\text{d}x} = -\frac{\text{d}\Phi}{\text{d}x} \tag{4.10}$$

where the discharge potential is

$$\Phi = kHh + C_c \qquad h \ge -z_b/\alpha \tag{4.11}$$

The constant C_c is chosen such that the discharge potential at the toe (where $h_{\text{toe}} = -z_b/\alpha$) is equal to Φ_{toe} (Eq. 4.9)

$$\Phi_{\text{toe}} = -kHz_b/\alpha + C_c = \frac{kH^2}{2\alpha} \tag{4.12}$$

which gives

$$C_c = \frac{\tfrac{1}{2}kH^2 + kHz_b}{\alpha} \tag{4.13}$$

With this choice for C_c, the potential is continuous when it transitions from regular confined flow to confined interface flow at the toe.

Now that steady confined interface flow is written as potential flow, any potential flow solution may be used that fulfills the differential equation for steady one-dimensional potential flow (Eq. 3.5)

$$\frac{\text{d}^2\Phi}{\text{d}x^2} = -N \tag{4.14}$$

Head-specified boundary conditions need to be transformed into discharge potentials using either Eq. (4.8) or (4.11), depending on the value of the head. Once a solution is obtained in terms of the discharge potential, the head may be computed by the inverse of Eq. (4.8) or (4.11)

$$h = \sqrt{\frac{2\Phi}{k\alpha}} - \frac{z_t}{\alpha} \qquad \Phi \leq \Phi_{\text{toe}} \tag{4.15}$$

$$h = \frac{\Phi - C_c}{kH} \qquad \Phi \geq \Phi_{\text{toe}} \tag{4.16}$$

Finally, the elevation z_i of the interface may be computed from the head as

$$z_i = -\alpha h \qquad h_{\text{tip}} \leq h \leq h_{\text{toe}} \tag{4.17}$$

As an example, consider uniform flow toward the coast at a rate $Q_x = U$ as shown in Figure 4.2 (Q_x is constant because the recharge $N = 0$). The origin of the coordinate system is chosen at the coastline, and the aquifer extends toward negative infinity. The elevation of the top of the confined aquifer is z_t with respect to the sea level. The head at the coastline is $-z_t/\alpha$, which is equivalent to an interface elevation of $z_i = z_t$. The corresponding discharge potential at the coastline is obtained by substituting $h = -z_t/\alpha$ in the equation for the discharge potential (Eq. 4.8), which gives $\Phi = 0$. Hence, the boundary conditions are

$$\Phi|_{x=0} = 0 \tag{4.18}$$

$$Q_x = -\frac{d\Phi}{dx} = U \tag{4.19}$$

The solution for the potential that fulfills the differential equation (4.14) with $N = 0$ and the boundary conditions is

$$\Phi = -Ux \tag{4.20}$$

The location of the toe is found at the position where $\Phi = \Phi_{\text{toe}}$, which gives

$$x_{\text{toe}} = -\frac{\Phi_{\text{toe}}}{U} = -\frac{kH^2}{2\alpha U} \tag{4.21}$$

Note that for a given value of U, the location of the toe is a function only of the aquifer parameters k and H, and the factor α, which is a function of the densities of freshwater and saltwater. The toe location is not a function of the sea level. And hence, the location of the toe remains where it is when the sea level rises, provided that the discharge toward the coast remains U. The heads in the aquifer increase with the sea level rise, of course.

In the example that follows, the location of the interface and the stream function are computed and plotted (Figure 4.3).

```
# parameters
k = 10 # hydraulic conductivity, m/d
zt = -10 # top of aquifer, m
zb = -30 # bottom of aquifer, m
rhof = 1000 # density of freshwater, kg/m^3
rhos = 1025 # density of saltwater, kg/m^3
U = 0.4 # flow toward the coast, m^2/d
H = zt - zb # aquifer thickness, m
```

```
# solution
alpha = rhof / (rhos - rhof)
phitoe = 0.5 * k * H ** 2 / alpha
Cc = (0.5 * k * H ** 2 + k * H * zb) / alpha
x = np.linspace(-200, 0, 150)
phi = -U * x
h = np.zeros_like(phi)
h[phi <= phitoe] = np.sqrt(2 * phi[phi <= phitoe] / (k * alpha)) - zt / alpha
h[phi > phitoe] = (phi[phi > phitoe] - Cc) / (k * H)
zi = np.maximum(-alpha * h, zb)
#
xtoe = -k * H**2 / (2 * alpha * U)
print(f'x-location of the toe: {xtoe:.0f} m')
# stream function
xg = [x, x]
zg = [zi, zt * np.ones(len(x))]
psi = [np.zeros(len(x)), -U * np.ones(len(x))]
```

```
x-location of the toe: -125 m
```

```
# basic streamline plot
plt.subplot(111, aspect=2)
plt.contour(xg, zg, psi)
plt.plot(x, zi, 'C1');
```

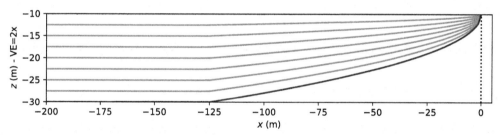

Figure 4.3 Streamlines in a vertical cross section for uniform flow toward the coast ($x = 0$) in a confined coastal aquifer.

The slopes of the streamlines change at the toe of the interface in Figure 4.3. This is a consequence of the Dupuit approximation. The tip of the interface is at the top of the aquifer at the coastline. This means that there is no gap for the freshwater to discharge into the sea. That is rather unrealistic and a consequence of adoption of the Dupuit approximation. For a coastline with a vertical beach face, as shown here with a dotted line, there is always a vertical outflow zone. The freshwater head varies along a vertical outflow zone, because the pressure distribution in the saltwater is hydrostatic, so that the freshwater head in the saltwater increases with depth. An example of two-dimensional interface flow in the vertical plane with an outflow zone along a horizontal sea bottom is presented in Section 10.5. Comparisons with interface solutions for two-dimensional flow in the vertical plane show that the location of the toe of the interface is often approximated reasonably well with a Dupuit solution, even though the depth of the interface near the coastline is underestimated (Sections 10.5 and 10.6).

Exercise 4.1: Redo the example for the case that the flow U is unknown, but the head is measured 200 m from the coastline and is equal to 1 m. Compute U and the location of the toe. Plot the head and interface vs. x.

Exercise 4.2: Redo the example for the case that the flow U is unknown, but the head is measured 200 m from the coastline and is equal to 0.5 m. Compute U and the location of the toe. Plot the head and interface vs. x.

Exercise 4.3: Reconsider the case of the example. Compute the location of the toe when the sea level rises by 0.5 m and all the other parameters remain the same. Compute the change in head in the aquifer.

Exercise 4.4: Reconsider the case of Exercise 4.1. Compute the location of the toe when the sea level rises by 0.5 m while the head 200 m from the coastline remains the same. Note that the head 200 m from the coastline is measured with respect to the original sea level. What is the effect of sea level rise in a confined aquifer when the inland boundary is a specified head?

4.2   Unconfined interface flow

Consider steady unconfined interface flow in a coastal aquifer as shown in Figure 4.4. The head and elevations are again measured with respect to sea level. The bottom of the aquifer is at $z = z_b$. As for confined flow, unconfined flow in coastal aquifers may consist of a part near the shore where the interface is present and a part of the aquifer farther inland where there is no interface. The problem is again solved in terms of a discharge potential in both parts of the aquifer.

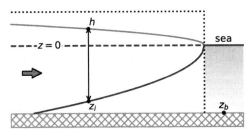

Figure 4.4 Steady interface flow in an unconfined aquifer.

First, consider the freshwater zone above the interface. The water table is a distance h above sea level, and the elevation of the interface is obtained with the Ghyben–Herzberg equation (4.4) as $z_i = -\alpha h$ so that the total thickness of the freshwater zone is equal to $h - z_i = (1+\alpha)h$. The discharge vector may be written as

$$Q_x = -k(\alpha + 1)h\frac{dh}{dx} = -\frac{d\Phi}{dx} \tag{4.22}$$

where the discharge potential for unconfined interface flow is

$$\Phi = \tfrac{1}{2}k(\alpha + 1)h^2 \qquad 0 \le h \le -z_b/\alpha \tag{4.23}$$

Second, consider the part of the flow upstream of the toe, where the flow is unconfined without an interface. The saturated thickness is equal to $h - z_b$ (recall that the head is measured with respect to the sea level) so that the discharge vector may be written as

$$Q_x = -k(h - z_b)\frac{dh}{dx} = -\frac{d\Phi}{dx} \tag{4.24}$$

where the discharge potential is

$$\Phi = \tfrac{1}{2}k(h - z_b)^2 + C_u \qquad h \ge -z_b/\alpha \tag{4.25}$$

The constant C_u is chosen such that the potential is continuous at the point where the flow changes from unconfined interface flow to regular unconfined flow, which is at the toe of the interface. The head at the toe is equal to $h_{\text{toe}} = -z_b/\alpha$, and the potential at the toe of the interface may be computed with Eq. (4.23) as

$$\Phi_{\text{toe}} = \tfrac{1}{2}k\frac{\alpha+1}{\alpha^2}z_b^2 \tag{4.26}$$

Constant C_u is obtained by setting the discharge potential (Eq. 4.25) equal to Φ_{toe}, with $h = -z_b/\alpha$ at the toe

$$\Phi_{\text{toe}} = \frac{1}{2}k\left(-\frac{z_b}{\alpha} - z_b\right)^2 + C_u = \frac{1}{2}k\frac{\alpha+1}{\alpha^2}z_b^2 \tag{4.27}$$

so that

$$C_u = -\tfrac{1}{2}k\frac{\alpha+1}{\alpha}z_b^2 \tag{4.28}$$

As for confined interface flow, head-specified boundary conditions need to be transformed into discharge potentials using either Eq. (4.23) or (4.25), depending on the value of the head. Once a solution is obtained in terms of the discharge potential, the head may be computed with the inverse of Eq. (4.23) or (4.25)

$$h = \sqrt{\frac{2\Phi}{k(\alpha+1)}} \qquad \Phi \leq \Phi_{\text{toe}} \tag{4.29}$$

$$h = \sqrt{\frac{2(\Phi - C_u)}{k}} + z_b \qquad \Phi \geq \Phi_{\text{toe}} \tag{4.30}$$

As an example, consider unconfined interface flow in a cross section of a long strip island. The distance between the two shores (i.e., the width of the island) is L, and the infiltration rate is equal to N. The origin of the coordinate system is chosen at the left coast (Figure 4.5). At both coasts, the head is equal to sea level

$$h|_{x=0} = h|_{x=L} = 0 \tag{4.31}$$

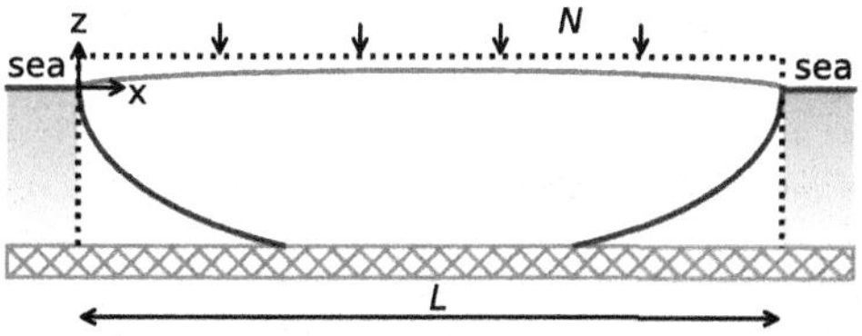

Figure 4.5 Unconfined interface flow below an infinite strip island with areal recharge.

The corresponding boundary conditions in terms of potentials are

$$\Phi|_{x=0} = \Phi|_{x=L} = 0 \tag{4.32}$$

The potential as a function of x for this case was derived in Section 3.3. Substitution of boundary conditions (4.32) in Eq. (3.39) gives

$$\Phi = -\frac{N}{2}(x^2 - Lx) \tag{4.33}$$

The interface reaches the bottom of the aquifer when the potential at the center of the island is larger than the potential at the toe Φ_{toe} (Eq. 4.26). In the example below, the solution is plotted for two values of the recharge N: one value for which the interface doesn't reach the bottom of the aquifer and one value for which the interface reaches the bottom of the aquifer (Figure 4.6).

```
# parameters
k = 40 # hydraulic conductivity, m/d
zb = -20 # bottom of aquifer, m
rhof = 1000 # density of freshwater, kg/m^3
rhos = 1025 # density of saltwater, kg/m^3
L = 1000 # width of island, m
N1 = 0.001 # recharge rate 1, m/d
N2 = 0.002 # recharge rate 2, m/d
alpha = rhof / (rhos - rhof) # alpha factor
```

```
# solution
def pot(x, N, L):
    phi = -0.5 * N * (x**2 - L * x)
    return phi

def pot2h(pot, k, alpha, zb):
    phitoe = 0.5 * k * (alpha + 1) / alpha**2 * zb**2
    Cu = -0.5 * k * (alpha + 1) * zb**2 / alpha
    h = np.zeros(len(pot))
    h[pot < phitoe] = np.sqrt(2 * pot[pot < phitoe] / (k * (alpha + 1)))
    h[pot >= phitoe] = np.sqrt(2 * (pot[pot >= phitoe] - Cu) / k) + zb
    return h

def zinterface(h, alpha, zb):
    return np.maximum(-alpha * h, zb)

x = np.linspace(0, L, 201)
phi1 = pot(x, N1, L)
phi2 = pot(x, N2, L)
h1 = pot2h(phi1, k, alpha, zb)
h2 = pot2h(phi2, k, alpha, zb)
zi1 = zinterface(h1, alpha, zb)
zi2 = zinterface(h2, alpha, zb)

phitoe = 0.5 * k * (alpha + 1) / alpha**2 * zb**2
print(f'Phi_toe: {phitoe:.2f}')
print(f'N={N1} results in maximum phi {np.max(phi1)}')
print(f'N={N2} results in maximum phi {np.max(phi2)}')
```

```
Phi_toe: 205.00
N=0.001 results in maximum phi 125.0
N=0.002 results in maximum phi 250.0
```

```
# basic plot
plt.subplot(111, aspect=10)
plt.plot(x, zi1, label=f'N={N1} m/d')
plt.plot(x, zi2, label=f'N={N2} m/d')
plt.legend();
```

When the recharge rate is large enough, the location of the toe of the interface may be obtained by setting $\Phi = \Phi_{\text{toe}}$ and solving for x, which gives

$$x_{\text{toe}} = \frac{L}{2} \pm \frac{1}{2}\sqrt{L^2 - \frac{8\Phi_{\text{toe}}}{N}} \qquad (4.34)$$

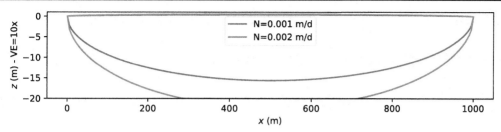

Figure 4.6   Unconfined interface flow in a cross section of an infinite strip island.

If the recharge is too small, the argument of the square root is negative and the interface does not reach the bottom of the aquifer. The location of the toe of the interface is a function of the sea level, because the equation for Φ_{toe} contains the elevation z_b of the base of the aquifer. Recall that z_b is measured with respect to the sea level. So when the sea level rises by 1 m, that means that z_b is 1 m lower than without the sea level rise.

Streamlines for the case with the larger recharge rate are shown below (Figure 4.7). As for confined flow, the simulated length of the outflow zone is zero, as a result of the Dupuit approximation. This same problem is solved for a horizontal sea floor without the Dupuit approximation in Section 10.6, which results in an outflow zone along the bottom of the sea.

```
# solution
xg = [x, x]
zg = [zi2, h2]
psi = [np.zeros_like(x), N2 * (x - L / 2)]
```

```
# basic streamline plot
plt.subplot(111, aspect=10)
plt.contour(xg, zg, psi, 20);
```

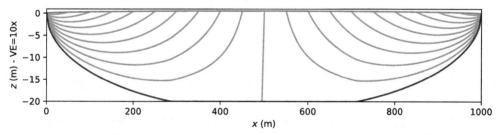

Figure 4.7   Streamlines for unconfined interface flow in a vertical cross section of an infinite strip island with recharge $N = 2$ mm/d.

Exercise 4.5: Compute the locations of the toes of the interface for the case that $N = 0.002$ m/d and all other parameters are as in the example. Next, compute the locations of the toes of the interface for the case that the sea level rises by 1 m.

Exercise 4.6: For the example above, compute the recharge N such that the interface exactly touches the bottom of the aquifer at $x = 500$ m. Plot the head and interface vs x.

Exercise 4.7: Compute and plot the elevation of the interface for the case of unconfined interface shown in Figure 4.8. Use $k = 20$ m/d, $N = 0.001$ m/d, $\rho_s = 1025$ kg/m³, $z_b = -40$ m, $L = 1000$ m. Compute the head at $x = 0$ and the location of the toe of the interface. Compare your answer for the head at $x = 0$ to the same problem of regular unconfined flow (all freshwater).

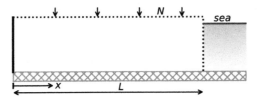

Figure 4.8 Situation for Exercise 4.7.

Exercise 4.8: Reconsider the case of Exercise 4.7. Compute the location of the toe when the sea level rises by 0.5 m and all the other parameters remain the same. Plot the head vs. x, using the original sea level as a datum. Compute the change in the position of the toe after the sea level rise.

4.3   Combined confined/semi-confined interface flow

Sections 4.1 and 4.2 dealt with interface flow to coastlines that were approximated as vertical. Here, the bottom of the sea is approximated as horizontal and separated from the aquifer by a leaky layer with resistance c (Figure 4.9). This means that flow in the part of the aquifer below the sea is semi-confined; the thickness of the leaky layer is neglected. As in Section 4.1, flow toward the coast is uniform at a rate $Q_x = U$, the origin of the coordinate system is chosen at the coast line, and the aquifer extends toward negative infinity. The top and bottom of the aquifer are z_t and z_b, respectively. Below the land, the aquifer is confined. The length of the outflow zone below the sea is unknown upfront. The sea bottom extends toward positive infinity (or at least beyond the point where the groundwater flow terminates).

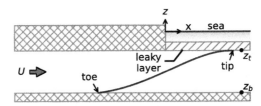

Figure 4.9 Interface flow toward a sea that is separated from the aquifer by a leaky layer.

The flow problem now consists of three parts: confined flow upstream of the interface toe, confined interface flow between the toe and the coast line, and semi-confined interface flow below the sea. The flow below the land can be solved with the discharge potential for confined interface flow derived in Section 4.1. The discharge potential cannot be used for the semi-confined flow section below the sea.

An expression for the flow in the semi-confined part is derived below. It is assumed that the toe of the interface is below the land (a condition is derived later). Just as for confined interface flow, the discharge vector below the sea can be written as (see Eq. 4.7)

$$Q_x = -k(\alpha h + z_t)\frac{dh}{dx} \qquad (4.35)$$

In the sea, the freshwater head h^* at elevation z_t is

$$h^* = -z_t/\alpha \qquad (4.36)$$

so that Q_x may be written as

$$Q_x = -k\alpha(h - h^*)\frac{dh}{dx} \qquad (4.37)$$

The flow below the sea is governed by the differential equation for semi-confined flow

$$\frac{dQ_x}{dx} = \frac{h - h^*}{c} \tag{4.38}$$

Substitution of Eq. (4.37) for Q_x in Eq. (4.38) and rearrangement of terms gives

$$\frac{d^2(h - h^*)^2}{dx^2} = \frac{2(h - h^*)}{k\alpha c} \tag{4.39}$$

The boundary conditions are that the discharge vector equals U at the coastline

$$Q_x|_{x=0} = U \tag{4.40}$$

and that the head equals h^* at the tip of the interface, which is at the a priori unknown location $x = x_{tip}$

$$h|_{x=x_{tip}} = h^* \tag{4.41}$$

The solution to Eq. (4.39) with boundary condition (4.41) is

$$h = h^* + \frac{(x - x_{tip})^2}{6k\alpha c} \qquad 0 \le x \le x_{tip} \tag{4.42}$$

It is easy to check this solution by substituting the function and its second derivative in the differential equation (4.39) and by evaluating the boundary conditions. An expression for the discharge vector is obtained from differentiation and substitution in Eq. (4.37), which gives

$$Q_x = -\frac{(x - x_{tip})^3}{18k\alpha c^2} \qquad 0 \le x \le x_{tip} \tag{4.43}$$

Finally, application of boundary condition (4.40) gives

$$x_{tip} = (18k\alpha c^2 U)^{1/3} \tag{4.44}$$

Note that this means that the outflow zone has a finite length and depends on the parameters of the aquifer and leaky layer, the density difference, and the total outflow into the sea. This is contrary to regular semi-confined flow (no interface), where the outflow zone is theoretically infinitely long and 95% of the water leaks through the semi-confining layer over a distance of three times the leakage factor ($\lambda = \sqrt{kHc}$), independent of the total outflow (Chapter 2).

The head h_0 at $x = 0$ is obtained from Eq. (4.42) as

$$h_0 = h^* + \frac{x_{tip}^2}{6k\alpha c} \tag{4.45}$$

Solution (4.42) is only valid if the toe of the interface is not below the sea, which means that

$$h_0 < -z_b/\alpha \tag{4.46}$$

Using that $z_b = z_t - H = -\alpha h^* - H$ and after some algebra, this gives the following condition for U

$$U \le \frac{kH^2\sqrt{2/3}}{\alpha\lambda} \tag{4.47}$$

The solution for the case that the interface toe is below the sea may be found in Bakker (2006).

Finally, a solution is obtained for the part of the aquifer below the land. The discharge potential at $x = 0$ is obtained from h_0 (Eq. 4.45) with the potential for confined interface flow (Eq. 4.8) as

$$\Phi_0 = \tfrac{1}{2}k\alpha(h_0 - h^*)^2 \tag{4.48}$$

The solution for the discharge potential below the land is

$$\Phi = -Ux + \Phi_0 \tag{4.49}$$

The toe of the interface is located at the point where $\Phi = \Phi_{\text{toe}}$ (Eq. 4.9), which gives

$$x_{\text{toe}} = -\frac{\Phi_{\text{toe}} - \Phi_0}{U} \tag{4.50}$$

In Section 4.1, the coast was approximated as vertical, which is equivalent to the current problem if the resistance of the leaky layer at the bottom of the sea is set to $c = 0$ in Eq. (4.45), which gives $h_0 = h^*$ (note that x_{tip} approaches zero for $c \to 0$). In the example below, the same parameters are used as in the example of Section 4.1, and the interface is computed for two values of the resistance of the leaky layer (Figure 4.10).

```
# parameters
k = 10 # hydraulic conductivity, m/d
zt = -10 # top of aquifer, m
zb = -30 # bottom of aquifer, m
alpha = 40 # alpha factor, -
clist = [1e-12, 5, 50] # three values of resistance of leaky layer, d
U = 0.4 # flow toward the coast, m^2/d
H = zt - zb # aquifer thickness, m
```

```
# solution
def interface(k, zt, zb, c, alpha, U):
    H = zt - zb
    # below sea
    hstar = -zt / alpha
    xtip = (18 * U * k * alpha * c**2) ** (1 / 3)
    xsea = np.linspace(0, xtip, 100)
    hsea = hstar + (xsea - xtip)**2 / (6 * k * alpha * c)
    h0 = hstar + xtip ** 2 / (6 * k * alpha * c)
    # below land
    phi0 = 0.5 * k * alpha * (h0 - hstar)**2
    phitoe = 0.5 * k * H ** 2 / alpha
    xtoe = -(phitoe - phi0) / U
    Cc = (0.5 * k * H ** 2 + k * H * zb) / alpha
    xland = np.linspace(xtoe, 0, 100)
    phi = -U * xland + phi0
    hland = np.sqrt(2 * phi/ (k * alpha)) + hstar
    # combine solution
    x = np.hstack((xland, xsea))
    zi = -alpha * np.hstack((hland, hsea))
    return x, zi
```

```
# basic plot
plt.subplot(111, aspect=4, ylim=(-30, -10))
for c in clist:
    x, zi = interface(k=k, zt=zt, zb=zb, c=c, alpha=alpha, U=U)
```

```
    plt.plot(x, zi, label=f'c={c:.0f} d')
plt.legend();
```

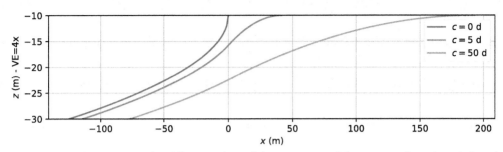

Figure 4.10 Interface elevation for different values of the resistance of the semi-confining layer below the
sea.

The streamlines for the case that $c = 5$ d are computed below and shown in Figure 4.11.

```
c = 5
x, zi = interface(k=k, zt=zt, zb=zb, c=5, alpha=alpha, U=U)
x = np.hstack((-200, x))
zi = np.hstack((zb, zi))
Qx = U * np.ones_like(x)
lab = np.sqrt(c * k * H)
xtip = (18 * U * k * alpha * c**2) ** (1 / 3)
Qx[x>0] = -(x[x>0] - xtip) ** 3 / (18 * k * alpha * c**2)
xg = [x, x]
zg = [zi, zt * np.ones_like(x)]
psi = [np.zeros_like(x), Qx]
```

```
# basic streamline plot
plt.subplot(111, aspect=2)
plt.contour(xg, zg, psi);
```

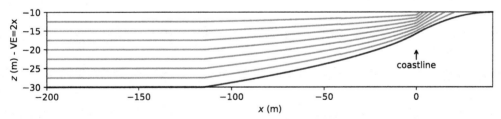

Figure 4.11 Streamlines in a vertical cross section for interface flow toward a sea that is separated from
the aquifer by a leaky layer.

Exercise 4.9: Redo the interface flow part of the example presented in Section 4.2, but now
the shore line is not vertical, but the sea bottom is horizontal at $z_t = 0$ on both sides of the
island and separated from the aquifer by a thin resistance layer with resistance $c = 5$ d; use
$N = 0.001$ m/d. Compute the head and interface elevation at the shore line ($x = 0$ or $x = L$)
and at the center of the island.

Chapter 5

Transient one-dimensional flow

All solutions presented so far in this book were for steady flow: The head and flow were a function of x, but not of time. In nature, all flows are transient and they vary in response to forcings such as rainfall, evaporation, transpiration, atmospheric pressure changes, tidal fluctuations, and geological processes (e.g., loading, tectonic stress, and heating). In addition to these natural processes, there is groundwater pumping and injection, which are anthropogenic forcings. The degree to which the approximation of steady state is justified depends on the system under consideration and the objective of the investigation. For example, groundwater fluctuations may be small, or fluctuations may not be of interest when the objective is to study the long-term average situation. It is up to the modeler to assess whether or not temporal fluctuations are of importance. Transient analytic solutions can be very useful for that purpose because they can be used to quickly determine how long it takes before a steady state is reached.

This chapter deals with one-dimensional transient flow, which means that the head and flow are a function of both x and time t. Transient solutions involving pumping are considered in Chapter 9. The focus here is on problems where heads vary as a result of changes in surface water levels or groundwater recharge.

As a starting point, the differential equation for one-dimensional transient flow is derived. The transient groundwater mass balance states that

$$\text{Mass in} - \text{Mass out} = \text{Increase in mass} \tag{5.1}$$

The groundwater mass balance reduces to a volume balance when the groundwater density is approximated as constant. The volume balance is

$$\text{Volume in} - \text{Volume out} = \text{Increase in volume} \tag{5.2}$$

The volume balance for one-dimensional flow in an aquifer is derived by considering a small section of aquifer that is Δx long and $\Delta y = 1$ wide in the direction normal to the plane of flow. Inflow consists of horizontal flow from the left, $Q_x(x, t)$ [L^2/T], and recharge at the top, N [L/T]. Outflow consists of horizontal flow at the right, $Q_x(x + \Delta x, t)$ (Figure 5.1). The increase in volume equals the increase in head multiplied by the storage coefficient S and $\Delta x \Delta y$ (where $\Delta y = 1$). Recall from Section 0.6 that the storage coefficient is equal to the phreatic storage $S = S_p$ for unconfined flow and the product of the specific storage S_s and the aquifer thickness H ($S = S_s H$) for confined flow. Further, recall that the discharge vector Q_x is the discharge per unit width of aquifer.

The volume balance is written for a period Δt. Substitution of the appropriate volumes in Eq. (5.2) gives

$$Q_x(x, t)\Delta t + N\Delta x \Delta t - Q_x(x + \Delta x, t)\Delta t = S\Delta x[h(x, t + \Delta t) - h(x, t)] \tag{5.3}$$

Division by $\Delta x \Delta t$ and rearrangement of terms gives

$$\frac{Q_x(x + \Delta x, t) - Q_x(x, t)}{\Delta x} = N - S\frac{h(x, t + \Delta t) - h(x, t)}{\Delta t} \tag{5.4}$$

DOI: 10.1201/9781315206134-5

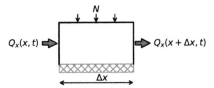

Figure 5.1 Transient water balance for a section of aquifer that is Δx long and $\Delta y = 1$ wide normal to the plane of flow.

In the limit for $\Delta x \to 0$ and $\Delta t \to 0$, the volume balance turns into the differential equation for transient one-dimensional horizontal transient flow

$$\frac{\partial Q_x}{\partial x} = -S\frac{\partial h}{\partial t} + N \tag{5.5}$$

The discharge vector for transient flow is the same as that for steady flow (Eq. 1.6)

$$Q_x = -T\frac{\partial h}{\partial x} \tag{5.6}$$

Recall that for confined flow, T is the constant transmissivity of the aquifer. For unconfined flow, the head-dependent transmissivity is approximated by a constant effective transmissivity $T = kH$, where the saturated thickness H is approximated as constant.

Substitution of Eq. (5.6) for Q_x in Eq. (5.5) gives

$$\frac{\partial^2 h}{\partial x^2} = \frac{S}{T}\frac{\partial h}{\partial t} - \frac{N}{T} \tag{5.7}$$

When the areal recharge equals zero, the governing differential equation reduces to

$$\frac{\partial^2 h}{\partial x^2} = \frac{S}{T}\frac{\partial h}{\partial t} \tag{5.8}$$

This equation is known as the diffusion equation; the term T/S is also referred to as the aquifer diffusivity. The diffusion equation governs the transient behavior of many other physical processes.

Solutions for transient groundwater flow are, not surprisingly, more difficult to obtain than those for steady flow. Common mathematical approaches include similarity solutions, separation of variables, Fourier series, and Laplace transforms. An introduction into many of these mathematical solution techniques for hydrological applications may be found in, e.g., Bruggeman (1999). In this chapter, solutions are presented without derivation. Verification of the solutions is discussed and part of the exercises. Solutions are presented for three problems: the response to a change in the surface water level in a semi-infinite aquifer (Section 5.1), the tidal response in a semi-infinite aquifer (Section 5.2), and the response to transient recharge between two parallel rivers (Section 5.3). The problem of Section 5.1 is resolved using the Laplace transform approach in Section 5.4, and a brief introduction into the (numerical) evaluation of Laplace transform solutions is presented. In the final section of this chapter, a solution to the Boussinesq equation for transient unconfined flow is presented.

5.1 Step changes in surface water level

Changes in river stage propagate as head changes in aquifers that are in contact with the river, which is important for understanding surface water–groundwater interactions. The following example considers one-dimensional transient flow in a semi-infinite aquifer as shown in

Figure 5.2. The aquifer is bounded on the left by a fully penetrating river that is in direct contact with the aquifer. Initially, the head in the aquifer is equal to the river stage h_0 everywhere. At time $t = t_0$, the stage in the river is raised by an amount Δh.

The solution is split into a steady part h_s that is only a function of x and a transient part h_t that is a function of x and t

$$h = h_s + h_t \tag{5.9}$$

The boundary conditions for the steady part are

$$h_s|_{x=0} = h_s|_{x \to \infty} = h_0 \tag{5.10}$$

The steady solution is easy: $h_s = h_0$.

The boundary conditions for the transient solution are

$$h_t|_{x=0, t>0} = \Delta h \tag{5.11}$$
$$h_t|_{x \to \infty, t} = 0 \tag{5.12}$$

and the initial condition is

$$h_t|_{x, t=0} = 0 \tag{5.13}$$

Note that the sum of the boundary conditions of the steady and transient solutions gives the boundary condition of the stated problem; the same holds for the initial condition. In essence, the transient solution represents the head change in the aquifer caused by the change in the river stage. The remainder of this section is concerned with the transient part of the solution, and the subscript $_t$ is dropped.

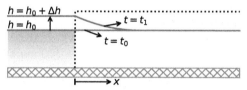

Figure 5.2 Response to an abrupt change in the surface water level ($t_1 > t_0$).

The solution for the stated transient problem was obtained as a similarity solution by Edelman (1947) and is referred to as the Edelman solution. Bruggeman (1999) obtained the same result using the Laplace transform method. The Edelman solution is

$$h(x, t) = \Delta h \, \mathrm{erfc}(u) \tag{5.14}$$

where

$$u = \sqrt{\frac{S x^2}{4T(t - t_0)}} \tag{5.15}$$

and erfc is the complimentary error function defined as

$$\mathrm{erfc}(u) = \int_u^\infty \frac{2}{\sqrt{\pi}} e^{-\tau^2} d\tau \tag{5.16}$$

The complementary error function is available in the `scipy.special` package.

An equation for the flux in the aquifer is obtained with Darcy's law

$$Q_x = -T \frac{\partial h}{\partial x} = -T \Delta h \frac{\partial \mathrm{erfc}(u)}{\partial x} \tag{5.17}$$

The derivative of the complimentary error function is obtained by first writing

$$\text{erfc}(u) = \int_u^\infty \frac{\mathrm{d}F}{\mathrm{d}\tau}\mathrm{d}\tau = F(\infty) - F(u) \tag{5.18}$$

where F is an (unknown) function with derivative

$$\frac{\mathrm{d}F}{\mathrm{d}\tau} = \frac{2}{\sqrt{\pi}}e^{-\tau^2} \tag{5.19}$$

The derivative of the complimentary error function may now be written as (note that $F(\infty)$ does not depend on u)

$$\frac{\partial \text{erfc}(u)}{\partial x} = -\frac{\mathrm{d}F}{\mathrm{d}u}\frac{\partial u}{\partial x} \tag{5.20}$$

The derivative $\partial u/\partial x$ is obtained by application of the chain rule

$$\frac{\partial u}{\partial x} = \frac{1}{2}\left(\frac{Sx^2}{4T(t-t_0)}\right)^{-1/2}\frac{2Sx}{4T(t-t_0)} = \frac{1}{u}\frac{u^2}{x} = \frac{u}{x} \tag{5.21}$$

so that finally

$$Q_x = T\Delta h \frac{2u}{x\sqrt{\pi}}e^{-u^2} \tag{5.22}$$

In the example below, the head and discharge vector are plotted as a function of x for three different times (Figure 5.3). The variation of the head with x shows where the additional water is stored in the aquifer. The amount of water that flows from the river into the aquifer decreases quickly, as can also be seen from the graph of Q_x. At $t = 1$ d, the inflow at $x = 0$ is more than 5 m²/d and decreases to almost zero at $x = 100$ m. At $t = 100$ d, the inflow at $x = 0$ has decreased to around 0.5 m²/d and is still around 0.5 m²/d at $x = 200$ m, which means that most water is carried beyond this point to increase storage in the aquifer.

```
# parameters
T = 100 # transmissivity, m^2/d
S = 0.2 # storage coefficient, -
delh = 2 # change in river level, m
t0 = 0 # time of change in river level, d
```

```
# solution
from scipy.special import erfc

def h_edelman(x, t, T, S, delh=1, t0=0):
    u = np.sqrt(S * x ** 2 / (4 * T * (t - t0)))
    return delh * erfc(u)

def Qx_edelman(x, t, T, S, delh, t0=0):
    u = np.sqrt(S * x ** 2 / (4 * T * (t - t0)))
    return T * delh * 2 * u / (x * np.sqrt(np.pi)) * np.exp(-u ** 2)
```

```
# basic plot head and Qx vs x
x = np.linspace(1e-12, 200, 100)
plt.subplot(121)
for t in [1, 10, 100]:
```

```
    h = h_edelman(x, t, T, S, delh, t0)
    plt.plot(x, h, label=f'time={t} d')
plt.legend()
plt.subplot(122)
for t in [1, 10, 100]:
    Qx = Qx_edelman(x, t, T, S, delh, t0)
    plt.plot(x, Qx, label=f'time={t} d')
plt.legend();
```

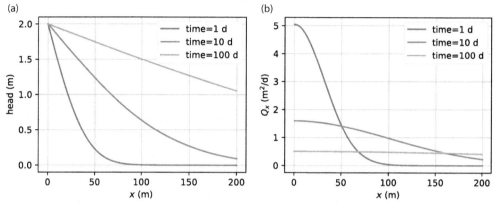

Figure 5.3 Snapshots of the groundwater response to an abrupt change in the surface water level; head (left) and Q_x (right).

In the following graphs, the head and Q_x are plotted as a function of t at three different distances from the river (Figure 5.4). The head as a function of t represents the head in an observation well. Note that the farther away an observation well is from the river, the longer it takes before the head starts to rise after the river stage is changed. The discharge vector Q_x shows a clear maximum near the river, but the peak becomes less pronounced as the distance to the river increases.

```
# basic plot head and Qx vs t
t = np.linspace(1e-12, 100, 100)
plt.subplot(121)
for x in [50, 100, 200]:
    h = h_edelman(x, t, T, S, delh, t0)
    plt.plot(t, h, label=f'distance {x} m')
plt.legend()
plt.subplot(122)
for x in [50, 100, 200]:
    Qx = Qx_edelman(x, t, T, S, delh, t0)
    plt.plot(t, Qx, label=f'distance {x} m')
plt.legend();
```

According to the solution, it takes infinite time before the head is raised by Δh everywhere in the aquifer. The head far away can only be raised if water flows from the river into the aquifer, which means that the head in the aquifer must be lower than the head in the river everywhere. This means that the gradient from the river to the point where the head is still near the initial head becomes smaller and smaller, so raising the head will go slower and slower. This is an artifact of the approximation that the aquifer extends to infinity, while in reality, some boundary will be present.

The solution for the head (Eq. 5.14) is only a function of the dimensionless parameter u, also referred to as the similarity variable (which combines the independent variables). That

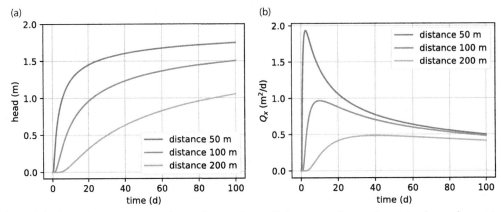

Figure 5.4 Groundwater response in an observation well due to an abrupt change in the surface water level at $t = 0$; head (left) and Q_x (right).

means that for one specific value of u, there is only one value for the head. For example, the head change at $x = 100$ m and $t = 10$ d may be computed as

```
x = 100 # m
t = 10 # d
T = 100 # m^2/d
S = 0.2 # -
print(f'head at x = {x} m, t = {t} d for T = {T} and S = {S}:')
print(f'u = {np.sqrt(S * x ** 2 / (4 * T * t)):.3f}')
print(f'h = {h_edelman(x, t, T, S, delh):.6f} m')
```

```
head at x = 100 m, t = 10 d for T = 100 and S = 0.2:
u = 0.707
h = 0.634621 m
```

If the storage coefficient is not $S = 0.2$ but $S = 0.002$, the same head change at $x = 100$ m is reached after 0.1 d, because $S = 0.002$ and $t = 0.1$ d give the same value of u.

```
t = 0.1 # d
S = 0.002 # -
print(f'head at x = {x} m, t = {t} d for T = {T} and S = {S}:')
print(f'u = {np.sqrt(S * x ** 2 / (4 * T * t)):.3f}')
print(f'h = {h_edelman(x, t, T, S, delh):.6f} m')
```

```
head at x = 100 m, t = 0.1 d for T = 100 and S = 0.002:
u = 0.707
h = 0.634621 m
```

Exercise 5.1: Consider the case when $T = 100$ m²/d, $S = 0.2$, and $\Delta h = 2$ m. The head at $x = 100$ m and $t = 10$ d is $h = 0.635$ m. For what value of the transmissivity is the head at $x = 100$ m and $t = 10$ d the same when $S = 0.002$ rather than $S = 0.2$?

Exercise 5.2: Consider the case when $T = 100$ m²/d, $S = 0.2$, and $\Delta h = 2$ m. The head at $x = 100$ m and $t = 10$ d is $h = 0.635$ m. At what distance from the river is the head change at $t = 10$ d the same when $S = 0.002$?

The accuracy and correct implementation of the solution can be verified in three ways. First, it can be verified that the mathematical solution (5.14) is indeed the solution to the stated mathematical problem by checking that the solution satisfies the boundary and initial conditions Eqs. (5.11)–(5.13), and by taking derivatives of the solution and substituting them in the governing differential equation (5.8) to check that the solution satisfies the differential equation. Second, the Python implementation of the solution may be verified by checking that the solution meets the differential equation and the boundary and initial conditions using accurate numerical derivatives (this approach was applied in Section 2.1).

Exercise 5.3: Verify that solution (5.14) meets the differential equation and the boundary and initial conditions, and verify that $Q_x = -T\mathrm{d}h/\mathrm{d}x$. Use accurate numerical derivatives (see, e.g., Section 2.1).

The third approach to verify the solution is to check the water balance. The total inflow into the aquifer over a period Δt can be computed and compared to the increase in the storage of water in the aquifer. The total inflow can be computed through integration of Q_x at $x = 0$. Q_x at $x = 0$ may be obtained from (5.22) and (5.15) as

$$Q_x|_{x=0,t} = \Delta h \sqrt{\frac{ST}{\pi(t - t_0)}} \tag{5.23}$$

The total inflow over a period Δt is

$$Q_{\mathrm{in}} = \int_{t_0}^{t_0 + \Delta t} Q_x|_{x=0,t} = 2\Delta h \sqrt{\frac{ST\Delta t}{\pi}} \tag{5.24}$$

The total increase in the storage may be computed through numerical integration along x of the head in the aquifer at $t = \Delta t$ and multiplication by the storage coefficient. The integration may be carried out, for example, by using the quad method of `scipy.integrate`. The total increase in storage must equal the total inflow, as is verified below.

```
delt = 10 # time period for which water balance is checked
Qin = 2 * delh * np.sqrt(S * T * delt / (np.pi))
from scipy.integrate import quad
stored = S * quad(h_edelman, 1e-12, np.infty, args=(delt, T, S, delh, 0))[0]
print(f'total inflow from river   : {Qin:.6f} m^3')
print(f'total increase in storage: {stored:.6f} m^3')
```

```
total inflow from river   : 3.191538 m^3
total increase in storage: 3.191538 m^3
```

As the governing differential equation (5.8) is linear, solutions may be superimposed through time. For example, if the water level in the river is raised by Δh_0 at time t_0 and then again by another Δh_1 at time t_1 (where $t_1 > t_0$), then the transient solution is

$$h(x, t) = \Delta h_0 \mathrm{erfc}(u_0) \qquad t_0 \leq t \leq t_1 \tag{5.25}$$
$$h(x, t) = \Delta h_0 \mathrm{erfc}(u_0) + \Delta h_1 \mathrm{erfc}(u_1) \qquad t \geq t_1 \tag{5.26}$$

where

$$u_0 = \sqrt{\frac{Sx^2}{4T(t - t_0)}} \qquad u_1 = \sqrt{\frac{Sx^2}{4T(t - t_1)}} \tag{5.27}$$

Consider the following sequence of four steps delh in the river level starting at four different times t0.

```
delh = np.array([1, 1, -2, 1]) # sequence of river stage changes, m
t0 = np.array([0, 10, 20, 30]) # sequence of times of changes, d
```

The transient solution is obtained through superposition of four changes in the river level. A loop is used to compute and plot the head vs. time at 10 and 100 m from the river (Figure 5.5).

```
# solution
t =
np.linspace(1e-3, 40, 400) # start just after t=0
h10 = np.zeros(len(t))
h100 = np.zeros(len(t))
for i in range(len(t0)):
    h10[t > t0[i]] += h_edelman(10, t[t > t0[i]], T, S, delh[i], t0[i])
    h100[t > t0[i]] += h_edelman(100, t[t > t0[i]], T, S, delh[i], t0[i])
```

```
# basic plot head (bottom graph)
plt.plot(t, h10, label='x=10 m')
plt.plot(t, h100, label='x=100 m')
plt.legend();
```

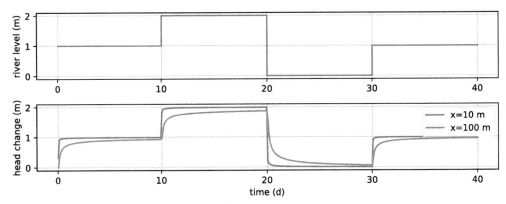

Figure 5.5 Head response due to a varying surface water level.

An interesting practical application is the analysis of the passage of a flood wave in the river. The river stage is raised by Δh for a period of Δt, after which the level goes back to the initial level. The head rise in the aquifer is obtained from superposition as

$$h(x, t) = \Delta h[\text{erfc}(u_0) - \text{erfc}(u_1)] \qquad t \geq \Delta t \tag{5.28}$$

where

$$u_0 = \sqrt{\frac{Sx^2}{4Tt}} \qquad u_1 = \sqrt{\frac{Sx^2}{4T(t - \Delta t)}} \tag{5.29}$$

The analysis of the groundwater response to a flood wave to estimate aquifer properties is sometimes referred to as the flood wave method. In this case, solution (5.28) may be fitted to estimate the aquifer diffusivity.

The groundwater response to a flood wave declines with the distance from the river. The distance from the river where the response is negligible (for example, a fraction of Δh) may be computed from solution (5.28), but that requires a numerical root-finding procedure. In practice, the distance where the maximum head response is a fraction α of the height of the flood wave may be estimated with

$$x \approx \sqrt{\frac{T\Delta t}{S\alpha}} \tag{5.30}$$

For example, a maximum response of $0.01 \Delta h$ ($\alpha = 0.01$) is reached at $x \approx 10\sqrt{T \Delta t / S}$. The time at which the maximum occurs is

$$t_{\text{peak}} = \frac{Sx^2}{6T} \tag{5.31}$$

The assessment of the accuracy of the approximate formula (5.30) is part of the exercises below.

Exercise 5.4: Consider a flood wave that passes through the river with $\Delta h = 2$ m and $\Delta t = 10$ d. Other parameters are as in the example. Use the approximate formula (5.30) to estimate the location where the head response is $0.01 \Delta h$. At this location, plot the head variation vs. time. Make sure that time varies from 0 to $5 t_{\text{peak}}$. Assess whether the approximate formula works reasonably well by computing the actual maximum head at the location computed with the approximate formula.

Exercise 5.5: For the flood wave of the previous exercise, compute and plot the total amount of water stored in the aquifer (on one side of the river) vs. time, for time going from 0 to $10 \Delta t$. How much water is stored in the river bank at $t = \Delta t = 10$ d? And how much at $t = 4 \Delta t = 40$ d?

5.2   Periodic changes in surface water level

Periodic head variations in an aquifer may be caused by, e.g., tides, seasonal fluctuations, or the periodic operation of dams. The surface water level variation is approximated here as a sinusoidal fluctuation with one specific amplitude and one specific period. Surface water level variations can commonly be expressed as a combination of several sinusoidal fluctuations with their own amplitude and period; the response to such a combined variation may be computed through superposition.

Consider one-dimensional flow in a semi-infinite aquifer (see Figure 5.6). There is no areal recharge and no flow at infinity

$$\frac{\partial h}{\partial x}\Big|_{x \to \infty, t} = 0 \tag{5.32}$$

The transmissivity T of the aquifer is approximated as constant. The aquifer is bounded by open water at $x = 0$, which fluctuates as

$$h|_{x=0,t} = A \cos\left(\frac{2\pi (t - t_p)}{\tau}\right) \tag{5.33}$$

where A is the amplitude of the fluctuation, t_p is the time of the peak, and τ is the period of the fluctuation. There is no initial condition; that is, the head at $x = 0$ has fluctuated as (5.33) forever (in practice, for a long time).

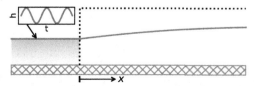

Figure 5.6   Groundwater response to periodic changes in surface water level.

The solution to the stated problem may be obtained with separation of variables and was first published by Forchheimer (1919, p. 1229). It turns out that the head response in the aquifer may be written in the same form as the head variation in the surface water, but the amplitude B of the variation is smaller than A and the peak arrives t_s later (Jacob, 1950, p. 365)

$$h(x, t) = B \cos[2\pi (t - t_p - t_s)/\tau] \tag{5.34}$$

The amplitude B of the head variation dampens with distance as

$$B = A \exp\left(-x\sqrt{\frac{S\pi}{T\tau}}\right) \tag{5.35}$$

and the time shift t_s of the peak is

$$t_s = x\sqrt{\frac{S\tau}{4\pi T}} \tag{5.36}$$

Note that both the amplitude B and the phase shift t_s are a function of the diffusivity T/S. This means that when the amplitude and peak time of a tidal response are measured to estimate aquifer parameters, only the aquifer diffusivity can be estimated, not the individual values of S or T. Estimation of (combinations of) aquifer parameters from the tidal response is also referred to as the tidal method (e.g., Ferris, 1952; Carr and Van Der Kamp, 1969).

Equation (5.35) shows that the amplitude B dampens out more quickly for shorter tidal periods than for longer tidal periods. The characteristic length μ is introduced as

$$\mu = \sqrt{\frac{T\tau}{S\pi}} \tag{5.37}$$

so that the amplitude may be written as

$$B = A e^{-x/\mu} \tag{5.38}$$

At a distance of $x = 3\mu$, the amplitude has dampened to 5% of the amplitude at $x = 0$. This result may be used as a rule of thumb to assess whether periodic fluctuations in surface water levels need to be taken into account when considering the head and flow in an aquifer. If the area of interest is farther away from a surface water body than 3μ, periodic fluctuations of the surface water level with a period of τ may be neglected.

As an example, consider an aquifer bounded by a tidal river. The head in the aquifer is plotted at four different times during one tidal period (Figure 5.7). The amplitude in the river is set to 1 m. The head variation is bounded by the amplitude B, which is shown with the dashed line.

```
# parameters
T = 1000 # transmissivity, m^2/d
S = 0.1 # storage coefficient, -
tau = 0.5 # tidal period, d
tp = 0 # time of peak at canal, d
mu = np.sqrt(T * tau / (S * np.pi)) # characteristic length, m
print(f'characteristic length: {mu:.2f} m')
```

```
characteristic length: 39.89 m
```

```
# solution for unit amplitude
def head(x, t, tau, S, T, tp=0, t0=0):
    B = np.exp(-x * np.sqrt(S * np.pi / (T * tau)))
```

```
    ts = x * np.sqrt(S * tau / (4 * np.pi * T))
    return B * np.cos(2 * np.pi * (t - tp - ts) / tau)
```

```
# basic plot
x = np.linspace(0, 200, 100)
for t in np.arange(0, tau, tau / 4):
    h = head(x, t, tau, S, T, tp)
    plt.plot(x, h)
plt.plot(x, np.exp(-x / mu), 'k--')
plt.plot(x, -np.exp(-x / mu), 'k--');
```

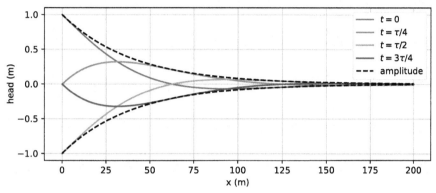

Figure 5.7 Tidal response in an aquifer at four different times.

Natural tidal fluctuations consist of a number of tidal components, each with its own period and phase lag. As an example, the response is presented to a tidal fluctuation consisting of two components. The first component is a semidiurnal component with a period of 745 minutes (referred to as component M2 in the system of tides); the amplitude of M2 is 1 m. The second component is a diurnal component with a period of 1549 minutes (referred to as component O1). The amplitude of O1 is 0.4 m, and the phase lag compared to M2 is 100 minutes. The solution is obtained through superposition of two solutions (5.34). The head fluctuation is computed at the shoreline and in three observation wells at three different distances from the coast line (Figure 5.8).

The presence of just two components causes the head fluctuation to become already quite complicated. Note that the limits of the vertical axes vary between the four graphs in Figure 5.8. The head fluctuates more than 2 m at the coast, while the variation at $x = 300$ m is only 5 mm. Component M2 damps out quicker than O2 as the period of M2 is smaller than the period of O2. The tidal fluctuation has two clear peaks near the coast, while there is only one clear peak farther away from the coast.

```
# parameters
tau1 = 745 / (24 * 60) # M2 period, d
tau2 = 1549 / (24 * 60) # O1 period, d
tp1 = 0 # time of peak M2, d
tp2 = 100 / (24 * 60) # time of peak of O1, d
mu1 = np.sqrt(T * tau1 / (S * np.pi)) # characteristic length M2, m
mu2 = np.sqrt(T * tau2 / (S * np.pi)) # characteristic length O1, m
print(f'characteristic length M2: {mu1:.2f} m')
print(f'characteristic length O1: {mu2:.2f} m')
```

```
characteristic length M2: 40.58 m
characteristic length O1: 58.52 m
```

```
# basic plot
t = np.linspace(0, 4, 400)
for i, x in enumerate([0, 100, 200, 300]):
    h = 1 * head(x, t, tau1, S, T, tp1) + \
        0.4 * head(x, t, tau2, S, T, tp2)
    plt.subplot(2, 2, i + 1)
    plt.plot(t, h)
```

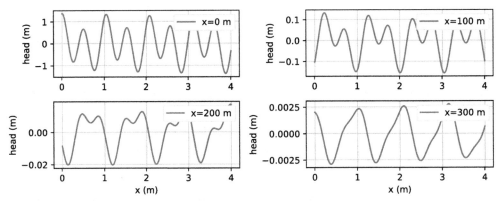

Figure 5.8 Tidal response to a tide consisting of two components; note the difference in vertical scale between the four graphs.

5.3   Areal recharge between two rivers

The solution for steady recharge between two parallel rivers in an aquifer with constant transmissivity T that was presented in Section 1.2 is extended to include transient recharge. The rivers are a distance L apart; the origin of the coordinate system is chosen halfway between the two rivers (Figure 5.9). Initially, the head in the aquifer is at steady state, so the initial transient head equals zero. The rivers are in direct contact with the aquifer and the river stages are fixed so that the boundary conditions and initial condition of the transient head are

$$h|_{x=-L/2,t} = h|_{x=L/2,t} = h|_{x,t=0} = 0 \tag{5.39}$$

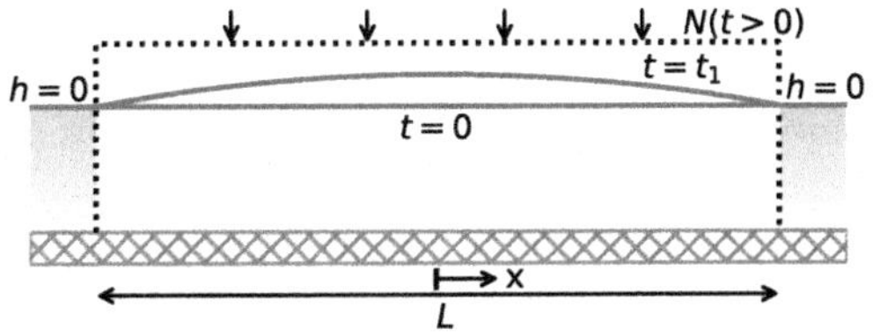

Figure 5.9 Transient recharge between two rivers ($t_1 > 0$).

At time $t = 0$, areal recharge starts at a constant and uniform rate N. The solution for the head in the aquifer as a function of time can be obtained with separation of variables and results in the following infinite series (Bruggeman, 1999, solution 133.16)

$$h = \frac{-N}{2T}\left(x^2 - \tfrac{1}{4}L^2\right) - \frac{4NL^2}{\pi^3 T}\sum_{n=0}^{\infty}\frac{(-1)^n}{(2n+1)^3}\cos\left[\frac{(2n+1)\pi x}{L}\right]\exp\left[-\frac{(2n+1)^2\pi^2 Tt}{SL^2}\right] \tag{5.40}$$

Note that for large times, the exponential term in the solution goes to zero and the head approaches the steady state solution $\frac{-N}{2T}\left(x^2 - \frac{1}{4}L^2\right)$, which is Eq. (1.24) for the case that the heads in the river are set to zero and the origin of the coordinate system is halfway between the two rivers.

In the following example, the head is plotted vs. x at three early times and at three late times (Figure 5.10). For the early times, the solution is plotted using three terms in the series (dashed lines), which is not sufficient as a distinct wiggle is visible, especially for $t = 1$ d. For later times, the dashed solution represents only one term in the series solution, which is indistinguishable from the solution with ten or more terms.

```
# parameters
L = 1000 # aquifer length, m
S = 0.1 # storage coefficient, -
T = 200 # transmissivity, m^2/d
N = 0.001 # recharge rate, m/d
```

```
# solution
def head(x, t, T=T, S=S, L=L, N=N, nterms=10):
    h = 0
    for n in range(nterms):
        h += (-1)**n / (2 * n + 1)**3 * \
        np.cos(((2 * n + 1) * np.pi * x) / L) * \
        np.exp(-((2 * n + 1)**2 * np.pi**2 * T * t) / (S * L**2))
    h = -N / (2 * T) * (x**2 - L**2/4) - 4 * N * L**2 / (np.pi**3 * T) * h
    return h
```

```
# basic head plot
plt.subplot(121) # early time
x = np.linspace(-L / 2, L / 2, 100)
for t in [1, 2, 3]:
    plt.plot(x, head(x, t, nterms=3), '--')
    plt.plot(x, head(x, t, nterms=10))
plt.subplot(122) # late time
for t in [40, 80, 160]:
    plt.plot(x, head(x, t, nterms=1) , '--') # just 1 term is enough
    plt.plot(x, head(x, t, nterms=10))
plt.plot(x, head(x, 0, nterms=0), 'k'); # 0 terms gives steady solution
```

At early times, the graph on the left shows that the head in the middle part of the aquifer rises at a rate $N/S = 0.01$ m/d, as discharge to the rivers does not have any draining effect on the middle part yet. When time progresses from 1 to 2 to 3 days, a larger and larger section of the aquifer drains toward the rivers. The head at $t = 160$ d approaches the steady state solution, as shown in the right graph.

The characteristic time t_{95} of the system is defined here as the time it takes for the head to reach 95% of the steady state level. An approximate equation for t_{95} is obtained by first approximating the solution (5.40) with the first term in the series solution

$$h \approx \frac{-N}{2T}\left(x^2 - \frac{1}{4}L^2\right) - \frac{4NL^2}{\pi^3 T}\cos\left[\frac{\pi x}{L}\right]e^{-t/\tau} \tag{5.41}$$

where

$$\tau = \frac{SL^2}{\pi^2 T} \tag{5.42}$$

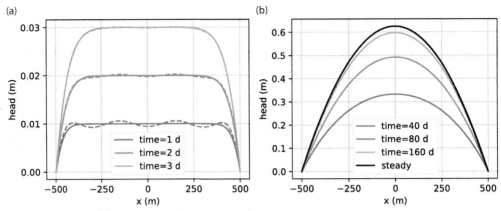

Figure 5.10 Transient infiltration between two rivers; early response with three terms (dashed) and ten terms (solid) (left) and late response with one term (dashed) and ten terms (solid) (right).

The cosine term may be further simplified with the first two terms of its Taylor series ($\cos(x) \approx 1 - x^2/2$)

$$-\frac{4NL^2}{\pi^3 T}\cos\left[\frac{\pi x}{L}\right] \approx -\frac{4NL^2}{\pi^3 T}\left(1 - \frac{\pi^2 x^2}{2L^2}\right) \approx \frac{N}{2T}\left(x^2 - \tfrac{1}{4}L^2\right) \tag{5.43}$$

where it is used that $\frac{4}{2\pi} \approx \frac{1}{2}$ and $\frac{8}{\pi^3} \approx \frac{1}{8}$. The head may now be approximated as

$$h \approx \frac{-N}{2T}\left(x^2 - \tfrac{1}{4}L^2\right)(1 - \mathrm{e}^{-t/\tau}) \tag{5.44}$$

It follows that at time $t = \tau$, the head has reached $1 - \mathrm{e}^{-1} = 63\%$ of its final value, and at time $t = 3\tau$, the head has reached 95% of its final value. Hence, the characteristic time t_{95} may be approximated as

$$t_{95} \approx \frac{3SL^2}{\pi^2 T} \tag{5.45}$$

Equation (5.45) is a reasonable approximation of t_{95} for the entire aquifer, as is explored in the exercise below. The characteristic time t_{95} is also referred to as the memory of the system, as the head is back to its original value t_{95} after a recharge event ends. For the parameters used here, the value of t_{95} may be computed as

```
t95 = 3 * S * L**2 / (np.pi**2 * T)
print(f'The memory of the system is: {t95:0.0f} d')
```

The memory of the system is: 152 d

Exercise 5.6: Compute the memory of the system t_{95} using 10 terms in the solution for 20 equally spaced points between the two rivers, and plot the memory of the system vs. x; use the fsolve method from scipy.optimize to find t_{95}, similar to the use of fsolve in Section 3.2. In the same graph, plot t_{95} according to Eq. (5.45).

Solution (5.40) is known as a step response: It is the response to a step in the recharge from zero to N. The unit step response is the response to a unit step in the recharge from zero to one. In addition to the unit step response, there is the unit block response, which is the response to a unit of recharge per day for a period Δt and zero after that; Δt is referred to as the length of the block. The block response b is obtained through superposition of two step responses

$$b(x, t) = s(x, t) \qquad 0 \le t \le \Delta t$$
$$b(x, t) = s(x, t) - s(x, t - \Delta t) \qquad t \ge \Delta t \tag{5.46}$$

where $s(x, t)$ is the unit step response starting at $t = 0$, and $s(x, t - \Delta t)$ is the unit step response starting at $t = \Delta t$. The unit step response $s(x, t)$ is computed with Eq. (5.40) by setting the recharge N to 1.

The step response and block response are plotted at two locations in Figure 5.11. This represents the head response in an observation well at the specified value of x. The block response is the same as the step response during the length of the block Δt. The dotted lines in the corresponding colors represent the final steady levels of the step responses. The black dashed line indicates the head rise in the absence of the rivers (i.e., the head rise is $N/S = 0.01$ m/d). The head in the observation well farthest away from the rivers (in this case, at $x = 0$) follows the dashed line for the longest period, up to ~20 days.

The memory of the system is indicated with the vertical lines representing the t_{95} times for the step response and the block response; the latter is obtained as t_{95} after the block recharge has ended. Note that for this case, the memory of the system is 152 days, and the head in the aquifer is approximately back to its original value 152 days after the end of the recharge event.

```
# plot step recharge and response and block recharge and response at x=0
x = 0
delt = 365 / 12
t = np.linspace(0, 210, 211)
plt.subplot(221)
plt.plot([0, 210], [1, 1])
plt.subplot(222)
plt.plot([0, 30, 30, 210], [1, 1, 0, 0])
plt.subplot(223)
h = N * head(x, t, N=1)
plt.plot(t, h)
plt.subplot(224)
h[t > delt] -= N * head(x, t[t > delt] - delt, N=1)
plt.plot(t, h);
```

The response to a series of recharge events of equal length Δt is computed using superposition. For example, if the recharge in the first three periods is N_0, N_1, and N_2, then the head at $t = 0, \Delta t, 2\Delta t$, and $3\Delta t$ can be computed as

$$h|_{x,0} = 0$$
$$h|_{x,\Delta t} = N_0 b|_{x,\Delta t}$$
$$h|_{x,2\Delta t} = N_0 b|_{x,2\Delta t} + N_1 b|_{x,\Delta t} \tag{5.47}$$
$$h|_{x,3\Delta t} = N_0 b|_{x,3\Delta t} + N_1 b|_{x,2\Delta t} + N_2 b|_{x,\Delta t}$$

A general equation for the head h_n at the beginning of period n is

$$h_n = \sum_{j=0}^{n-1} N_j b_{n-1-j} \tag{5.48}$$

where $h_n = h|_{x,n\Delta t}$ and $b_{n-1-j} = b|_{x,(n-1-j)\Delta t}$. Implementation of the summation in Python requires extra attention because in mathematical notation, the upper limit in a summation is included, while in Python, the upper limit is excluded.

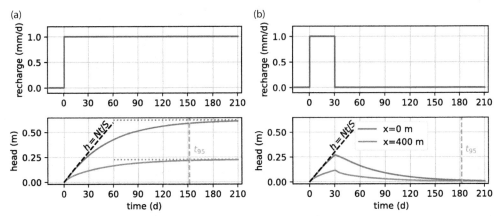

Figure 5.11 Step recharge (upper left) and the resulting step response with t_{95} after the start of step (lower left); block recharge for 30 days (upper right) and the resulting step response with t_{95} after the end of block (lower right); black dashed line is $h = Nt/S$.

As an example, consider the following four years of recharge values (in mm/d) for monthly periods (the recharge is constant during each month). These recharge values are from an area with a temperate climate with a shallow water table. The recharge is highest in the fall and winter and can become negative in the spring and summer months when vegetation draws water from the water table.

```
# four years of monthly recharge data
delt = 365 / 12 # approximate length of 1 month, d
recharge = 0.001 * np.array( # convert recharge from mm/d to m/d
[ 1.2,   1.5,   0.8,  -1.2,   0.2,  -2.9,  -0.6,   2.8,   1.2,   0.9,   3.8,
  1.8,   1.9,   1.8,  -0.6,  -1.3,  -2.5,   0.6,   0.9,   1.8,  -0.3,   0.8,
 -0.2,   3.8,   2.8,   0.3,  -0.5,   0.3,  -1.1,   1.5,   1. ,  -1.5,   0.1,
  1.5,   1. ,   4. ,   1.3,   1.3,   0.3,  -1.1,   0.6,  -0.8,  -1.2,  -1.6,
  1.3,   1.4,   2.4,   2.6])
ntime = len(recharge) # number of months
```

A function is written for the discrete block response $b(x, n, \Delta t)$, which is the head response for a unit of recharge during a block of length Δt. The discrete block response returns the head at ntime equal intervals Δt.

```
# discrete block response
def bresponse(x, ntime, delt, T=T, S=S, L=L, nterms=10):
    t = np.linspace(delt, ntime * delt, ntime)
    return head(x, t, T, S, L, 1, nterms) - \
           head(x, t - delt, T, S, L, 1, nterms)
```

The head is computed and plotted at two points in the aquifer through the application of Eq. (5.48) in Figure 5.12. Note that the unit block response is computed only once at each location. The yearly head variation halfway between the two rivers ($x = 0$) is almost 2 m, while the head variation closer to the river is much smaller. The highest heads occur around January, and the lowest heads around July. The computed head in the first months (shown by a dashed line) is not very realistic as it is based on recharge since the beginning of the first year only, while the head also depends on the recharge in the previous year. After $t_{95} = 152$ days (the memory of this system), the effect of the previous year is negligible.

```
# solution
h0 = np.zeros(ntime + 1) # head at x=0
h400 = np.zeros(ntime + 1) # head at x=400
b0 = bresponse(0, ntime, delt, T, S, L)
b400 = bresponse(400, ntime, delt, T, S, L)
for n in range(1, len(recharge) + 1):
    h0[n] = np.sum(recharge[0:n] * b0[n - 1::-1])
    h400[n] = np.sum(recharge[0:n] * b400[n - 1::-1])
```

```
# basic plot recharge data and head response
plt.subplot(211)
time = np.linspace(0, ntime * delt, ntime + 1)
plt.step(time[:-1], 1000 * recharge, where='post'); # plt.stairs in mpl>=3.4
plt.grid()
plt.subplot(212)
plt.plot(time, h0, label='$x=0$ m')
plt.plot(time, h400, label='$x=400$ m')
plt.legend();
```

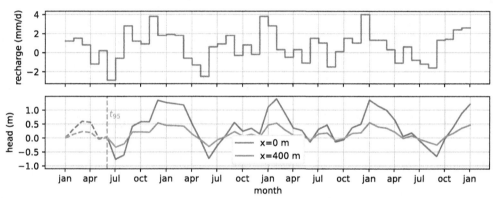

Figure 5.12 Four years of monthly varying recharge (top) and the resulting transient head variation for recharge between two rivers; the head is dashed for part of series before $t = t_{95}$.

The head shown in Figure 5.12 is computed with Eq. (5.48), which is based on the exact solution (5.40). As a final example, it is investigated whether the head variation at $x = 0$ shown in Figure 5.12 can be simulated accurately with the simplified step response function

$$s(t) = A \left(1 - e^{-t/a}\right) \tag{5.49}$$

The functional form of the simplified step response is the same as Eq. (5.44), but the parameters A and a are now simply fitting parameters. In other words, the head at $x = 0$ shown in Figure 5.12 is treated as the observed head in an observation well and the parameters A and a are obtained by performing a least squares fit. First, a function is written for the discrete block response corresponding to the simplified step response (5.49).

```
def simplebresponse(A, a, delt, nstep):
    t = np.linspace(0, nstep * delt, nstep + 1)
    hstep = A * (1 - np.exp(-t / a))
    return hstep[1:] - hstep[:-1]
```

The difference ε between the observed head h_o and the modeled head h_m, also called the residual or error, is computed as

$$\varepsilon = h_o - h_m \tag{5.50}$$

The optimal parameters A and a are found by minimizing the sum of squared errors, which is referred to as the objective function

$$F_{\text{obj}} = \sum \varepsilon^2 \tag{5.51}$$

The first argument of the Python function for the objective function must be an array with the parameters. The final keyword argument is `return_heads`. When this keyword argument is set to `True`, the method returns the modeled heads rather than the sum of the squared errors, which is useful to compute the modeled heads after the parameters are estimated.

```
def fobj(p, delt=delt, recharge=recharge, hobs=h0, return_heads=False):
    A, a = p
    ntime = len(recharge)
    b = simplebresponse(A, a, delt, ntime + 1)
    hm = np.zeros(ntime + 1)
    for n in range(1, ntime + 1):
        hm[n] = np.sum(recharge[0:n] * b[n - 1::-1])
    if return_heads:
        return hm
    rv = np.sum((hobs - hm) ** 2)
    return rv
```

The parameters that minimize the objective function may be found with a number of methods. Here, the `fmin` function of the `scipy.optimize` package is used, which applies the simplex algorithm (see the `fmin` documentation).

```
from scipy.optimize import fmin
A, a = fmin(fobj, [300, 50], disp=0)
print(f'Estimated parameters A: {A:.2f} d, a: {a:.2f} d')
hm = fobj([A, a], return_heads=True) # compute modeled heads
```

```
Estimated parameters A: 629.89 d, a: 53.03 d
```

The modeled heads h_m fit the observed heads h_o very well (Figure 5.13).

```
# basic plot modeled heads
plt.plot(h0, 'C1.', label='observed')
plt.plot(hm, 'C0', label='modeled')
plt.legend();
```

The procedure discussed above, where a model of heads in an observation well is created with an approximate response function, is known as time series analysis. Time series analysis is a powerful and increasingly popular method to analyze heads measured in observation wells. See, for example, Bakker and Schaars (2019) and Collenteur et al. (2019).

Exercise 5.7: Create a synthetic data set by computing 4 years of monthly heads at $x = 250$ m using the exact solution and the provided 4 years of monthly recharge. Apply time series analysis on the synthetic data set with the exponential response function (5.49). Report the fitted values of A and a. Plot the modeled head and observed (synthetic) head in one graph.

Figure 5.13 Fit of model with approximate response function to observations obtained with the exact solution for the response to four years of monthly varying recharge.

5.4   Solutions with Laplace transforms

As mentioned, there are various mathematical approaches to obtain solutions for transient groundwater flow problems. One of these approaches is the Laplace transform, which was originally applied to derive, e.g., the solution in Section 5.1. In the Laplace transform method, the governing partial differential equation is transformed into an ordinary differential equation in what is called the Laplace domain. The solution to the ordinary differential equation in the Laplace domain is much easier than the solution to the original partial differential equation. The difficulty lies in the back-transformation from the Laplace domain to the time domain, which requires the solution of a complicated integral. Luckily, there are several algorithms to carry out the integration numerically when analytic integration is cumbersome or impossible. There are entire books written on Laplace transformations, including useful tables with transformations of mathematical functions. A nice introduction for the mathematically inclined groundwater engineer or hydrogeologist is given by Bruggeman (1999). Here, a very brief introduction is given by demonstrating how to solve the problem of Section 5.1 for the case that the initial head equals zero.

The Laplace transform is an integral transform, indicated with the symbol $\mathcal{L}$, and is defined as

$$\mathcal{L}\{h\} = \bar{h}(x, p) = \int_0^\infty h(x, t)\mathrm{e}^{-pt}\mathrm{d}t \tag{5.52}$$

where $\bar{h}$ is the Laplace-transformed head, which is a function of x and the Laplace transform parameter p, but not of time. The transforms of derivatives with respect to x are simple

$$\mathcal{L}\left\{\frac{\partial h}{\partial x}\right\} = \frac{\partial \bar{h}}{\partial x} \qquad \mathcal{L}\left\{\frac{\partial^2 h}{\partial x^2}\right\} = \frac{\partial^2 \bar{h}}{\partial x^2} \tag{5.53}$$

while transformation of the derivative with respect to time is

$$\mathcal{L}\left\{\frac{\partial h}{\partial t}\right\} = p\bar{h} \tag{5.54}$$

Application of the Laplace transformation rules to the governing differential equation (5.8) for transient one-dimensional flow without recharge gives

$$\frac{\mathrm{d}^2\bar{h}}{\mathrm{d}x^2} = \frac{pS}{T}\bar{h} \tag{5.55}$$

which is indeed an ordinary differential equation. Eq. (5.55) is the modified Helmholtz equation (2.5), which was introduced in Chapter 2 for steady semi-confined flow.

Boundary conditions need to be transformed as well. Transformations of all kinds of functions can be found in tables of Laplace transforms. The Laplace transform of the constant Δh gives

$$\mathcal{L}\{\Delta h\} = \int_0^\infty \Delta h e^{-pt}dt = \frac{\Delta h}{p} \tag{5.56}$$

The solution to differential equation (5.55) with transformed boundary conditions results in an equation for $\bar{h}(x, p)$; this solution is referred to as the Laplace domain solution. The ultimate interest is of course a function $h(x, t)$, which is obtained through what is called the back-transformation or the inverse Laplace transformation. Several approaches exist to find the inverse transform (e.g., Bruggeman, 1999), but they are often difficult or even impossible to carry out analytically. This has spurred the development of several approximate algorithms to perform the back-transformation. The Stehfest algorithm is one of the most popular methods and is used here (Stehfest, 1970).

The Stehfest approximation of the inverse transformation is a weighted mean of the solution in the Laplace domain for a number of values of the Laplace parameter p_k ($k = 1, ..., M$). Hence, the application of the Stehfest approximation requires evaluation of $\bar{h}(x, p)$ for a number of values of the Laplace parameter p to compute the value of the head at location x and time t. The number of terms in the approximation is M, which must be even. The Stehfest formula is

$$h(x, t) = \frac{\ln 2}{t} \sum_{k=1}^{M} V_k \bar{h}(x, p_k) \tag{5.57}$$

where M is the number of terms, V_k is the weight for term k, and p_k is the Laplace parameter k computed as

$$p_k = \frac{k \ln 2}{t} \tag{5.58}$$

In theory, the approximation is more accurate when more terms are used, but machine accuracy is quickly causing a problem. The summation in Eq. (5.57) consists of the summation of increasingly larger positive and negative numbers. For large values of the index k in the summation, these numbers are so large that they only differ beyond the 16th digit. Python normally only keeps track of 16 digits, which means that the sum of a positive and negative number that only differ beyond the 16th digit essentially results in garbage. Luckily, it is fairly obvious when too many terms are used, as the result is obviously wrong.

The formula for the parameters V_k is rather lengthy

$$V_k = (-1)^{k+M/2} \sum_{j=\text{int}[(k+1)/2]}^{\min(k,M/2)} \frac{j^{M/2}(2j)!}{(\frac{M}{2} - j)!j!(j - 1)!(k - j)!(2j - k)!} \tag{5.59}$$

The formula for V_k is implemented in the routine below (adopted from the mpmath package by Johansson et al., 2013).

```
# coefficients V_k
from scipy.special import factorial as fac
def stehfest_coef(M):
    assert M % 2 == 0, 'M must be even' # make sure M is even
    M2 = int(M / 2)
    V = np.zeros(M)
    for k in range(1, M + 1):
        z = np.zeros(min(k, M2) + 1)
```

```
        for j in range(int((k + 1) / 2), min(k, M2) + 1):
            z[j] = j ** M2 * fac(2 * j) / \
            (fac(M2 - j) * fac(j) * fac(j - 1) * fac(k - j) * fac(2 * j - k))
        V[k - 1] = (-1) ** (k + M2) * np.sum(z)
    return V
```

The Stehfest algorithm (5.57) is implemented in the routine below. The first two input arguments are the x locations and times t for which the function is computed. Both input arguments x and t may be arrays. The next input argument is a Python implementation of the Laplace-transformed function that can be called as func(p, x, keyword_arguments), where keyword_arguments can be an arbitrary number of keyword arguments. The next input argument is the keyword argument $M = 12$, which is the default number of terms used in the Stehfest algorithm. Finally, an arbitrary number of keyword arguments may be supplied. These keyword arguments are passed to the function func of the Laplace-transformed solution (the keyword_arguments mentioned above). The stehfest function returns a two-dimensional array with shape (len(x), len(t)). Note that the parameters V_k are only a function of M, so they are computed only once. The Laplace parameters p_k must be computed separately for each time with Eq. (5.58).

```
# Stehfest algorithm
def stehfest(x, t, func, M=12, **kwargs):
    t = np.atleast_1d(t)
    x = np.atleast_1d(x)
    f = np.zeros((len(x), len(t)))
    V = stehfest_coef(M)
    for i in range(len(t)):
        p = np.arange(1, M + 1) * np.log(2) / t[i]
        for j in range(len(x)):
            fbar = func(p, x[j], **kwargs)
            f[j, i] = np.log(2) / t[i] * np.sum(V * fbar)
    return f
```

In the code blocks that follow, the transient problem of Section 5.1 (the groundwater response to a change Δh in the river stage in a semi-infinite confined aquifer) is resolved using Laplace transforms. The differential equation in the Laplace domain is (5.55), and the boundary conditions are

$$\bar{h}|_{x=0} = \frac{\Delta h}{p} \tag{5.60}$$

$$\bar{h}|_{x\to\infty} = 0$$

Mathematically, this is the same as the problem solved in Section 2.1. The solution is

$$\bar{h} = \frac{\Delta h}{p} \exp(-x\sqrt{pS/T}) \tag{5.61}$$

The solution in the time domain is obtained with the Stehfest algorithm, and the Stehfest solution is compared to the exact solution.

```
# parameters
T = 100 # transmissivity, m^2/d
S = 0.2 # storage coefficient, -
delh = 2 # increase in river level, m
```

```
# Laplace-transformed solution
def hbar(p, x, T=100, S=1e-3, delh=2):
    return delh / p * np.exp(-x * np.sqrt(p * S / T))
```

```
# comparison
x = 50 # location, m
t = 10 # time, d
M = 12 # number of terms
V = stehfest_coef(M)
hx = h_edelman(x, t, T, S, delh=2) # from Section 5.1
hs = stehfest(x, t, hbar, M=M, T=T, S=S)
print(f'Exact head at x={x} m and t={t} d: {hx}')
print(f'Stehfest head with M={M} terms  : {hs}')
print(f'Relative error: {(hs - hx) / hx}')
```

```
Exact head at x=50 m and t=10 d: 1.2341501549039475
Stehfest head with M=12 terms  : [[1.23415675]]
Relative error: [[5.34721333e-06]]
```

The relative error is on the order of 10^{-6} with $M = 12$ terms in the Stehfest solution. The relative error is plotted versus the number of Stehfest terms in Figure 5.14. Note that the relative error starts to increase again when more than 20 terms are used; this many terms should be avoided.

```
Mlist = np.arange(2, 33, 2)
error = np.zeros(len(Mlist))
for i, M in enumerate(Mlist):
    hs = stehfest(x, t, hbar, M=M, T=T, S=S)
    error[i] = np.abs((hs - hx) / hx)
```

```
# simple plot
plt.semilogy(Mlist, error, '-', marker='.')
plt.xticks(np.arange(2, 33, 2));
```

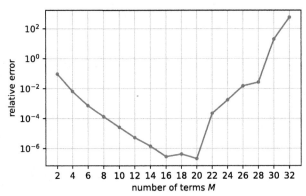

Figure 5.14   Relative error of the Stehfest solution as a function of the number of terms M used in the Stehfest algorithm.

The lowest relative error is on the order of 10^{-6}. This means that the Stehfest algorithm can be expected to be accurate to up to about six significant digits. The head is computed and plotted vs. x at three different times (Figure 5.15). The error is shown in the right graph of Figure 5.15.

```
# solution
x = np.linspace(1e-12, 200, 100)
t = [1, 10, 100]
hx = np.empty((len(x), 3))
for itime in range(3):
```

```
    hx[:, itime] = h_edelman(x, t[itime], T, S, delh) # h_edelman from Sec. 2.1
hs = stehfest(x, t, hbar, M=12, T=T, S=S)
```

```
# basic plot head and error
plt.subplot(121)
plt.plot(x, hs)
plt.subplot(122)
plt.plot(x, hs - hx);
```

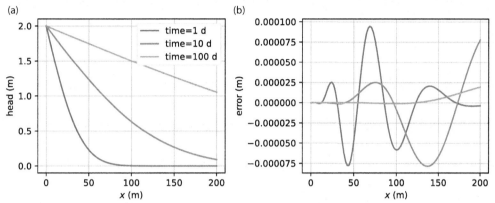

Figure 5.15 Stehfest solution for transient flow caused by a change in river stage; head (left) and Stehfest solution minus Edelman solution (right).

Exercise 5.8: Solve the transient flow problem of Section 5.1 with the Laplace transform method, but for the case that the aquifer is semi-confined. Use $T = 100$ m²/d, $S = 2 \cdot 10^{-4}$, and $c = 1000$ d for the resistance of the leaky layer. Reproduce the left graph of Figure 5.14 for x varying from 0 till 500 and $t = 0.01, 0.1,$ and 1 day; plot the solution for both the confined and semi-confined cases in the same graph. Report the head at $x = 100$ m at $t = 0.01$, 0.1, and 1 day.

5.5   Unconfined flow with variable transmissivity

So far in this chapter, the transmissivity of an aquifer has been approximated as constant, even when the flow was unconfined. This is a reasonable approximation when the variation of the saturated thickness is not too large, as was shown in Chapter 3 for steady unconfined flow. In this section, an alternative formulation for transient unconfined flow is presented and the accuracy of the constant transmissivity approximation is assessed.

The one-dimensional continuity equation for transient unconfined flow is given by Eq. (5.5) as

$$\frac{\partial Q_x}{\partial x} = -S_p \frac{\partial h}{\partial t} + N \tag{5.62}$$

where S_p is the phreatic storage (or specific yield). Q_x may be written in terms of the discharge potential for unconfined flow as Eq. (3.3)

$$Q_x = -\frac{d\Phi}{dx} \tag{5.63}$$

where Φ is given by Eq. (3.2)

$$\Phi = \tfrac{1}{2}k(h - z_b)^2 \tag{5.64}$$

where z_b is the elevation of the bottom of the aquifer. Substitution of Eq. (5.63) for Q_x in Eq. (5.62) gives

$$\frac{\partial^2 \Phi}{\partial x^2} = S \frac{\partial h}{\partial t} - N \tag{5.65}$$

This differential equation is known as the Boussinesq equation and is a nonlinear differential equation in terms of the head, because Φ includes a term h^2. This equation is notoriously difficult to solve analytically, even though it may not look all that complicated at first sight.

The differential equation has to be linearized to solve it analytically. First, consider the derivative of Φ (Eq. 5.64) with respect to time

$$\frac{\partial \Phi}{\partial t} = k(h - z_b) \frac{\partial h}{\partial t} \tag{5.66}$$

Differential equation (5.65) may now be written as

$$\frac{\partial^2 \Phi}{\partial x^2} = \frac{S}{k(h - z_b)} \frac{\partial \Phi}{\partial t} - N \tag{5.67}$$

This equation is linearized by approximating the saturated thickness in front of the time derivative by a constant value $\tilde{H}$

$$\frac{\partial^2 \Phi}{\partial x^2} \approx \frac{S}{k\tilde{H}} \frac{\partial \Phi}{\partial t} - N \tag{5.68}$$

Differential equation (5.68) is a linear differential equation in terms of Φ. The solution for the discharge potential may be written as the sum of a steady solution and a transient solution.

The problem of Section 5.1 is repeated, but now using the linearized Boussinesq equation for unconfined transient flow (Eq. 5.68). A semi-infinite aquifer is bounded by a canal at $x = 0$. Initially, the head is equal to h_0 so that the initial steady discharge potential equals

$$\Phi|_{x,t=0} = \Phi_0 = \tfrac{1}{2}k(h_0 - z_b)^2 \tag{5.69}$$

At time $t = 0$, the head in the canal is raised by Δh (Figure 5.2) so that the discharge potential increases by an amount $\Delta \Phi$

$$\Delta \Phi = \tfrac{1}{2}k(h_0 + \Delta h - z_b)^2 - \Phi_0 \tag{5.70}$$

The boundary condition of the transient solution at $x = 0$ is

$$\Phi|_{x=0,t>0} = \Delta \Phi \tag{5.71}$$

while the head at infinity remains equal to h_0

$$\Phi|_{x \to \infty, t} = 0 \tag{5.72}$$

The transient solution for the discharge potential is (compare Eq. 5.14)

$$\Phi(x, t) = \Delta \Phi \operatorname{erfc}(u) \tag{5.73}$$

where u is

$$u = \sqrt{\frac{S_p x^2}{4k\tilde{H}t}} \tag{5.74}$$

The solution for the head is obtained from the total discharge potential (the steady solution plus the transient solution) in the regular way using Eq. (3.7) as

$$h = z_b + \sqrt{\Delta\Phi\,\mathrm{erfc}(u) + \Phi_0} \qquad (5.75)$$

The example from Section 5.1 is resolved below. The linearized saturated thickness in the storage term is approximated as $\tilde{H} = h_0 + \frac{1}{2}\Delta h$. The solution to the linearized Boussinesq equation is compared to the solution with constant transmissivity (5.14) of Section 5.1 using $T = k\tilde{H}$ (Figure 5.16). For the numbers used in the example, the saturated thickness at $x = 0$ increases by 20%. The comparison shows only a small difference between the heads computed with the two methods, which means that the effect of a 20% increase in the saturated thickness is small. The difference in head computed with the two methods increases when the relative change in the saturated thickness is more dramatic, as is explored in the exercise below.

```
# parameters
k = 10 # hydraulic conductivity, m
zb = 0 # aquifer base, m
Sp = 0.2 # phreatic storage coefficient, -
h0 = 10 # initial head and canal level, m
delh = 2 # change in canal level, m
Ht = h0 + 0.5 * delh # linearized saturated thickness, m
```

```
from scipy.special import erfc
def h_unconfined(x, t, k, H, S, zb, h0, delh):
    pot0 = 0.5 * k * (h0 - zb) ** 2
    delpot = 0.5 * k * (h0 + delh - zb) ** 2 - pot0
    u = np.sqrt(S * x ** 2 / (4 * k * H * t))
    pot = delpot * erfc(u) + pot0
    return zb + np.sqrt(2 * pot / k)
```

```
# basic plot head
x = np.linspace(1e-12, 400, 100)
for t in [1, 10, 100]:
    h = h_edelman(x, t, k * Ht, Sp, delh) + h0 # from Section 5.1
    plt.plot(x, h)
    hun = h_unconfined(x, t, k, Ht, Sp, zb, h0, delh)
    plt.plot(x, hun)
```

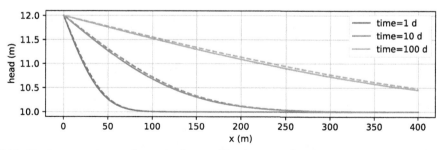

Figure 5.16 Head response to a change in the canal level at $x = 0$ using a constant transmissivity (solid lines) and a linearized saturated thickness in the storage term (dashed lines).

Exercise 5.9: Compare the solution for unconfined transient flow presented here with the solution of Section 5.1 for the case that the head in the canal suddenly drops by 5 m. Compare the head in the aquifer after $t = 1, 10, 100$ days. Does this drastic (possibly unrealistic) change in the saturated thickness result in a significant difference between the two solutions?

Chapter 6

Steady two-dimensional flow to wells

Many three-dimensional groundwater flow problems can be approximated as two-dimensional in the horizontal plane through adoption of the Dupuit–Forchheimer approximation. Of particular interest are problems that include wells. In this chapter, solutions are presented for extraction wells, injection wells, and wells near rivers and impermeable boundaries. Different solution techniques are demonstrated, depending on the type of problem. For homogeneous confined and unconfined aquifers, solutions are formulated in terms of a discharge potential so that the same potential solution can be used for confined flow, unconfined flow, and interface flow. The stream function is used in the horizontal plane for problems without areal infiltration. Solutions for flow to wells in semi-confined aquifers and two-aquifer systems are formulated in terms of heads.

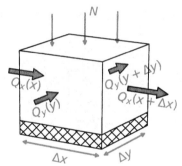

Figure 6.1 Two-dimensional water balance.

The discharge vector for two-dimensional flow in the horizontal plane consists of two components: Q_x in the x-direction and Q_y in the y-direction (Figure 6.1). Continuity of flow for steady two-dimensional flow in the horizontal x, y plane may be written as

$$\frac{\partial Q_x}{\partial x} + \frac{\partial Q_y}{\partial y} = N \qquad (6.1)$$

where N is the areal recharge. Note that this equation differs from the continuity of flow equation for one-dimensional steady flow (Eq. 1.5) only by the additional term $\partial Q_y / \partial y$.

Steady confined, unconfined, and interface flows may be formulated as potential flow, where the discharge vector is minus the gradient of the discharge potential

$$Q_x = -\frac{\partial \Phi}{\partial x} \qquad Q_y = -\frac{\partial \Phi}{\partial y} \qquad (6.2)$$

so that the differential equation for steady potential flow becomes the two-dimensional Poisson equation

$$\nabla^2 \Phi = \frac{\partial^2 \Phi}{\partial x^2} + \frac{\partial^2 \Phi}{\partial y^2} = -N \qquad (6.3)$$

DOI: 10.1201/9781315206134-6

Table 6.1   Definition of the discharge potential for different types of flow

Flow Type	Discharge potential	Condition	Section
Confined flow	$\Phi = kHh$	$h \geq z_t$	3.3
Unconfined flow	$\Phi = \frac{1}{2}k(h - z_b)^2$	$h \geq z_b$	3.1
Confined/uncon-	$\Phi = kHh - (\frac{1}{2}kH^2 + kHz_b)$	$h \geq z_t$	3.3
fined flow	$\Phi = \frac{1}{2}k(h - z_b)^2$	$z_b \leq h \leq z_t$	3.3
Confined interface	$\Phi = kHh + (\frac{1}{2}kH^2 + kHz_b)/\alpha$	$h \geq -z_b/\alpha$	4.1
flow	$\Phi = \frac{1}{2}k\alpha(h + z_t/\alpha)^2$	$-z_t/\alpha \leq h \leq -z_b/\alpha$	4.1
Unconfined inter-	$\Phi = \frac{1}{2}k(h - z_b)^2 - \frac{1}{2}k(\alpha + 1)z_b^2/\alpha$	$h \geq -z_b/\alpha$	4.2
face flow	$\Phi = \frac{1}{2}k(\alpha + 1)h^2$	$0 \leq h \leq -z_b/\alpha$	4.2

where k is the hydraulic conductivity of the aquifer, z_b and z_t are the elevations of the bottom and top of the aquifer, H is the aquifer thickness, and $\alpha = \rho_f/(\rho_s - \rho_f)$, where ρ_f and ρ_s are the densities of freshwater and saltwater, respectively. For interface flow, all heads and elevations are measured with respect to the sea level.

In the absence of areal recharge ($N = 0$), this differential equation is referred to as the two-dimensional Laplace equation. The relationships between the discharge potential and head for different flow types are summarized in Table 6.1.

Flow to a pumping well in a homogeneous, isotropic aquifer is radially symmetric in the absence of any other flow features. As a result, the head and flow are a function only of the radial distance to the well. The continuity equation in terms of radial coordinates is derived by considering the water balance for an annulus (Figure 6.2). The inner radius of the annulus is r, the outer radius is $r + \Delta r$, and Q_r is the radial component of the discharge vector (i.e., positive away from the well). For steady flow, the continuity equation is

$$\text{Inflow} - \text{Outflow} = 0 \tag{6.4}$$

Inflow equals areal recharge on the annulus and inflow along the inner radius. Outflow equals outflow along the outer radius so that the continuity equation becomes

$$(\pi(r + \Delta r)^2 - \pi r^2)N + 2\pi r Q_r|_r - 2\pi(r + \Delta r)Q_r|_{r+\Delta r} = 0 \tag{6.5}$$

Division by $2\pi \Delta r$ and taking the limit for $\Delta r \to 0$ gives

$$Nr - \frac{d(rQ_r)}{dr} = 0 \tag{6.6}$$

Application of the product rule for differentiation and using that the radial component of the discharge vector is

$$Q_r = -\frac{d\Phi}{dr} \tag{6.7}$$

gives the differential equation for steady potential flow in radial coordinates

$$\frac{d^2\Phi}{dr^2} + \frac{1}{r}\frac{d\Phi}{dr} = -N \tag{6.8}$$

Note that this equation is consistent with Eq. (6.3) because the Laplacian in terms of polar coordinates r, θ is

$$\nabla^2\Phi = \frac{d^2\Phi}{dr^2} + \frac{1}{r}\frac{d\Phi}{dr} + \frac{1}{r^2}\frac{d^2\Phi}{d\theta^2} \tag{6.9}$$

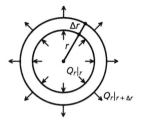

Figure 6.2 Continuity of flow for radially symmetric flow.

In radial coordinates, there is no dependence on θ so that the derivative with respect to θ drops out.

In Chapters 1–4, the stream function was computed in the vertical plane for one-dimensional Dupuit–Forchheimer flow. Recharge entered the vertical plane through the boundaries, and flow in the vertical plane was divergence-free (Eq. 1.28). The lower case symbol ψ [L^2/T] was used for the stream function in the vertical plane.

The stream function also exists for steady two-dimensional Dupuit flow in the horizontal plane if the flow is divergence-free. In the horizontal plane, flow is divergence-free if the areal recharge N equals zero

$$\frac{\partial Q_x}{\partial x} + \frac{\partial Q_y}{\partial y} = 0 \tag{6.10}$$

so that the discharge potential is governed by Laplace's equation

$$\nabla^2 \Phi = 0 \tag{6.11}$$

In the horizontal plane, the upper case symbol Ψ is used for the stream function, which has dimensions L^3/T.

For two-dimensional flow in the horizontal plane, the components of the discharge vector can be obtained from the stream function, as is explained in the following. Recall that the stream function is defined by three characteristics: The stream function is a function that is constant along streamlines, the difference of the stream function values between two points is equal to the amount of water flowing between the two points, and the value of the stream function increases to the right when looking in the direction of flow. Consider points A, B, and C shown in Figure 6.3. Water flows from left to right across AB. The component of the discharge vector normal to AB is Q_x, the distance between A and B is Δy, and the difference in the stream function between A and B is $\Delta \Psi$ so that

$$Q_x \Delta y = \Delta \Psi = \Psi(x, y) - \Psi(x, y + \Delta y) \tag{6.12}$$

hence

$$Q_x = -\frac{\Psi(x, y + \Delta y) - \Psi(x, y)}{\Delta y} \tag{6.13}$$

In the limit for Δy to zero, this gives

$$Q_x = -\frac{\partial \Psi}{\partial y} \tag{6.14}$$

Similarly, water flows from bottom to top across BC. The component of the discharge vector normal to BC is Q_y, the distance between B and C is Δx, and the difference in the stream function between B and C is $\Delta \Psi$, so that

$$Q_y \Delta x = \Delta \Psi = \Psi(x + \Delta x, y) - \Psi(x, y) \tag{6.15}$$

and hence

$$Q_y = \frac{\partial \Psi}{\partial x} \tag{6.16}$$

Comparison of Eqs. (6.14) and (6.16) with Eq. (6.2) gives

$$\frac{\partial \Phi}{\partial x} = \frac{\partial \Psi}{\partial y}$$
$$\frac{\partial \Phi}{\partial y} = -\frac{\partial \Psi}{\partial x} \tag{6.17}$$

This set of two equations is known as the Cauchy–Riemann conditions. Since the discharge potential and stream function fulfill the Cauchy–Riemann conditions, they can be combined into a complex potential, as is explained in Section 7.3.

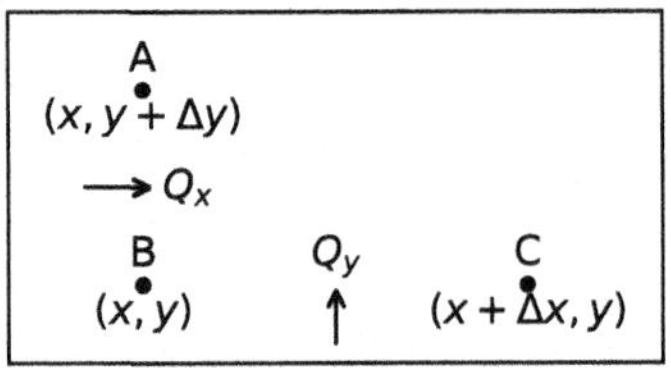

Figure 6.3 Flow between points A–B and B–C used in the derivation of the stream function.

The discharge potential is a single-valued function, which means that the discharge potential has only one value at a point. A mathematical definition of a single-valued function is (e.g., Strack, 1989)

$$\frac{\partial^2 \Phi}{\partial x \partial y} = \frac{\partial^2 \Phi}{\partial y \partial x} \tag{6.18}$$

Application of the Cauchy–Riemann conditions (6.17) to replace Φ by Ψ in Eq. (6.18) gives

$$\frac{\partial^2 \Psi}{\partial x^2} + \frac{\partial^2 \Psi}{\partial y^2} = 0 \tag{6.19}$$

This is a nice result, because it means that the stream function is governed by Laplace's equation, just like the discharge potential (Eq. 6.11). Recall that the stream function in the horizontal plane only exists when the areal recharge N equals zero.

6.1   Radially symmetric flow on a circular island

Radial flow problems and solutions are important in groundwater engineering because they represent the local flow field around pumping wells. In this section, solutions are presented for radially symmetric two-dimensional flow on a circular island. The head is fixed along the perimeter of the island, the areal recharge on the island is uniform, and a pumping well is located at the center of the island. The head and flow are a function only of the radial coordinate r. The problem is depicted for unconfined flow in Figure 6.4, but is equally valid for any of the flow types listed in Table 6.1. The solution is derived as follows. First, the solution is derived for flow to a well at the center of a circular island without recharge ($N = 0$). Second, the solution is derived for flow on a circular island with uniform recharge, but no well ($Q = 0$). And third, the first two solutions are combined to solve the problem for flow to a well at the center of a circular island with uniform recharge. The purpose of this solution approach is to demonstrate the principle of superposition.

The differential equation for the discharge potential in terms of r is given by Eq. (6.8). The general solution to this differential equation is

$$\Phi = -\frac{1}{4}Nr^2 + A\ln r + B \tag{6.20}$$

where A and B are constants to be evaluated from boundary conditions. The corresponding radial component of the discharge vector is

$$Q_r = -\frac{d\Phi}{dr} = \frac{1}{2}Nr - \frac{A}{r} \tag{6.21}$$

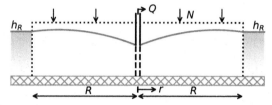

Figure 6.4 Unconfined flow to a well at the center of a circular island with radius R and uniform recharge.

First, consider flow to a pumping well at the center of a circular island of radius R; the radius of the well is r_w. The discharge potential that is the solution for this problem is referred to as Φ_1. The head is fixed to h_R at the perimeter of the island ($r = R$). The corresponding discharge potential is computed with the appropriate equation from Table 6.1 and is referred to as Φ_R

$$\Phi_1|_{r=R} = \Phi_R \tag{6.22}$$

The discharge of the well is Q [L³/T] and is positive for extracting water from the aquifer. The discharge is equal to the total flow into the well, which is the product of the discharge vector Q_r (Eq. 6.21) at the perimeter of the well and the circumference of the well, resulting in the boundary condition

$$Q = -2\pi r_w Q_{r_1}|_{r=r_w} \tag{6.23}$$

Note the minus sign, as Q_r is defined as positive in the positive r direction, and the discharge Q of the well is positive for pumping water out of the aquifer. A slightly simpler boundary condition is

$$Q = \lim_{r_w \to 0} -2\pi r_w Q_{r_1}|_{r=r_w} \tag{6.24}$$

This condition often produces accurate results as the radius of the well is commonly much smaller than the horizontal scale of the groundwater problem being considered. Note that for this special case of a well at the center of a circular island without areal recharge, Eq. (6.23) is valid for any r, not just at r_w.

Application of boundary conditions (6.22) and (6.24) to general solution (6.20), using $N = 0$, results in the solution for the discharge potential

$$\Phi_1 = \frac{Q}{2\pi}\ln(r/R) + \Phi_R \tag{6.25}$$

This is the potential form of the well-known Thiem equation (Thiem, 1906). The corresponding radial component of the discharge vector is

$$Q_{r_1} = -\frac{d\Phi_1}{dr} = -\frac{Q}{2\pi}\frac{1}{r} \tag{6.26}$$

Note that Q has the dimensions of a discharge (e.g., m^3/d), while Q_r is the radial component of the discharge vector with dimensions of discharge per unit length (e.g., m^2/d).

Neither the discharge potential (and thus the head), nor the discharge vector can be calculated at the well, where $r = 0$. In mathematical terms, such a point is called a singularity. The discharge potential (Eq. 6.25), the corresponding head, and the discharge vector (Eq. 6.26) have no physical meaning at the center of the well. These equations were derived for flow in the aquifer, but the aquifer stops at the well screen. Inside the well, the head is equal to the head in the aquifer just outside the well screen, provided the resistance to flow of the well screen is negligible. The head inside the well is computed by evaluating the head at the radius of the well ($r = r_w$).

Next, consider uniform recharge on the island and no pumping well; this solution is referred to as Φ_2. As in the previous solution, the head (and thus the potential) is fixed at the perimeter of the island

$$\Phi_2|_{r=R} = \Phi_R \tag{6.27}$$

The second boundary condition is obtained from symmetry: The gradient of the head is zero at the center of the island, and hence, the flow is zero

$$Q_{r_2}|_{r=0} = 0 \tag{6.28}$$

Application of these boundary conditions to the general solution (6.20) results in

$$\Phi_2 = -\frac{1}{4}N(r^2 - R^2) + \Phi_R \tag{6.29}$$

The total amount of groundwater flow in the aquifer at the perimeter of the island is equal to the total recharge on the island, as is to be expected from continuity of flow

$$2\pi R Q_{r_2}|_{r=R} = N\pi R^2 \tag{6.30}$$

The flow at the perimeter is outflow, which is opposite to the first solution, in which flow at the perimeter is inflow.

Finally, consider both a pumping well with discharge Q at the center of the island and uniform recharge N on the island. This problem contains both features of the first two solutions: recharge and a pumping well. The solution is referred to as Φ_3. The boundary conditions are

$$\Phi_3|_{r=R} = \Phi_R \tag{6.31}$$

and

$$\lim_{r_w \to 0} -2\pi r_w Q_{r_3}|_{r=r_w} = Q \tag{6.32}$$

Application of the boundary conditions to the general solution yields the following discharge potential

$$\Phi_3 = -\frac{1}{4}N(r^2 - R^2) + \frac{Q}{2\pi}\ln(r/R) + \Phi_R \tag{6.33}$$

and the corresponding discharge vector

$$Q_{r_3} = \frac{Nr}{2} - \frac{Q}{2\pi}\frac{1}{r} \tag{6.34}$$

Comparison of the three solutions Φ_1, Φ_2, and Φ_3 (Eqs. 6.25, 6.29, and 6.33) reveals that the third solution is the sum of the first two solutions except for the additive constant. Addition of multiple solutions to obtain another solution is an example of the principle of superposition, which is applicable to all linear differential equations, including the equations of Laplace and Poisson. The sum of the potentials of the first two problems satisfies the differential equation of the third problem

$$\nabla^2\Phi_3 = \nabla^2(\Phi_1 + \Phi_2) = \nabla^2\Phi_1 + \nabla^2\Phi_2 = 0 - N = -N \tag{6.35}$$

Similarly, the boundary condition at the well is satisfied by the sum of the two potentials. Finally, the value of the sum of the two potentials at $r = R$ is a constant

$$\Phi_1|_{r=R} = \Phi_2|_{r=R} = 2\Phi_R \tag{6.36}$$

This is not the value specified in the boundary condition ($\Phi_3|_{r=R} = \Phi_R$), but this is easily corrected by modification of the additive constant. In this example of superposition, two radially symmetric solutions were added such that the resulting solution is also two-dimensional radially symmetric flow, because the radial symmetry is about the same point (the center of the island).

The flow rate and direction at the island's perimeter depend on the relative magnitudes of the well discharge and the total recharge on the island. As an example, the head is computed for the case of unconfined flow in the code below. Two solutions are shown: the case for which the well pumps half the total areal recharge on the island and the case for which the well pumps exactly all the areal recharge on the island (Figure 6.5).

```
# parameters
k = 5 # hydraulic conductivity, m/d
zb = 0 # bottom elevation of aquifer, m
N = 0.001 # areal recharge, m/d
R = 200 # radius of the island, m
hR = 10   # head at edge of island, m
rw = 0.3 # well radius, m
Qhalf = 0.5 * N * np.pi * R ** 2   # discharge is half total recharge, m^3/d
Qtotal = N * np.pi * R ** 2   # discharge equals total recharge, m^3/d
```

```
# solution
r = np.linspace(rw, R, 400)
PhiR = 0.5 * k * (hR - zb) ** 2
Phi2 = -0.25 * N * (r ** 2 - R ** 2) + PhiR
Phi3 = Phi2 + Qhalf / (2 * np.pi) * np.log(r / R)
hhalf = zb + np.sqrt(2 * Phi3 / k)
Phi3 = Phi2 + Qtotal / (2 * np.pi) * np.log(r / R)
htotal = zb + np.sqrt(2 * Phi3 / k)
```

```
# basic plot head for half total discharge
plt.plot(-r, hhalf, 'C0', label='Q = half total recharge')
plt.plot(r, hhalf, 'C0')
plt.xticks(np.arange(-200, 201, 50), np.abs(np.arange(-200, 201, 50)))
plt.legend();
```

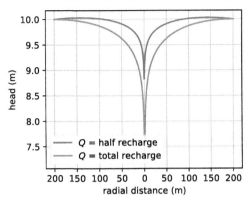

Figure 6.5 Head along centerline of a circular island for unconfined flow with uniform recharge and pumping well at the center of the island.

As expected, the drawdown is much larger for the latter case than for the former case. Note that for the latter case, the flow at the perimeter of the island is zero and thus the slope of the water table is zero there. In the former case, there is still outflow across the island perimeter. Inflow occurs along the perimeter once the well discharge exceeds the total recharge on the island. The maximum discharge of the well is reached when the water level in the well (at $r = r_w$) is at the bottom of the aquifer.

Exercise 6.1: Consider the case of a well at the center of a circular island with recharge where the discharge of the well is equal to half the total recharge on the island, using the example values above. Compute the distance r_d from the center of the well to the water divide by setting $Q_r|_{r=r_d} = 0$. Compare the total recharge on a circle with radius r_d to the discharge of the well.

Exercise 6.2: Consider the case of a well at the center of a circular island with recharge where the discharge of the well is equal to half the total recharge on the island, using the example values above. Compute the total flow into the well at $r = r_w$ by integrating Q_r around the well radius, and compare your answer to the discharge of the well. Is boundary condition (6.24) a good approximation?

Exercise 6.3: For the example given above of a well at the center of a circular unconfined island with recharge, compute the maximum discharge of the well by setting the head at the well equal to the bottom of the aquifer.

6.2   Wells near rivers and impermeable boundaries

In the previous section, the superposition principle was applied to obtain the solution for a well pumping at the center of a circular island with recharge. Here, solutions are presented for pumping wells near a surface water boundary (e.g., river, stream, canal, or lake) and an impermeable boundary. All boundaries are approximated as long and straight; areal recharge is not simulated. Solutions for the discharge potential, the discharge vector, and the stream function are obtained through superposition and the method of images.

The general solution for the discharge potential of well number n with discharge Q_n is obtained from the Thiem equation (6.25) by removing the constant part $\Phi_R - \frac{Q}{2\pi}\ln(R)$ (an appropriate constant is added later depending on the boundary condition)

$$\Phi = \frac{Q_n}{2\pi}\ln(r_n) = \frac{Q_n}{4\pi}\ln(r_n^2) \tag{6.37}$$

where r_n is the radial distance from point (x_n, y_n) to well n

$$r_n = \sqrt{(x - x_n)^2 + (y - y_n)^2} \tag{6.38}$$

The x- and y-components of the discharge vector of well n are obtained as

$$Q_x = -\frac{\partial \Phi}{\partial x} = -\frac{Q_n}{2\pi} \frac{x - x_n}{r_n^2} \tag{6.39}$$

$$Q_y = -\frac{\partial \Phi}{\partial y} = -\frac{Q_n}{2\pi} \frac{y - y_n}{r_n^2} \tag{6.40}$$

Exercise 6.4: A well is pumping at an unknown, steady rate in a confined aquifer with transmissivity $T = 1200$ m^2/day. Near the pumping well are two observation wells. Observation well A is located 8 m from the pumping well, and observation well B is located 24 m from the pumping well. The heads in these wells are $h_A = 134.2$ m and $h_B = 134.3$ m. Assuming that without the well pumping there is no hydraulic gradient in the aquifer, what is the pumping rate of the well?

The stream function for a well is derived next. A single well in a homogeneous, isotropic aquifer of infinite extent results in circular equipotentials. Streamlines are straight lines to the well. As such, the stream function is a function of θ only, where θ is the angle measured in counterclockwise direction with respect to the positive horizontal axis in a local coordinate system centered at the well (see the left graph of Figure 6.6). The stream function for well n is

$$\Psi = \frac{Q_n}{2\pi} \theta_n \tag{6.41}$$

where

$$\theta_n = \arctan\left(\frac{y - y_n}{x - x_n}\right) \tag{6.42}$$

where θ is measured in radians and not degrees.

Eight streamlines are plotted in Figure 6.6. The total discharge between each two consecutive streamlines is $Q/8$. The angle θ is measured with respect to the positive x-axis and jumps from π to $-\pi$ along the negative x-axis when going in the counterclockwise (the mathematically positive) direction. As a result, the stream function jumps across the negative x-axis, from $Q/2$ to $-Q/2$. This jump is unavoidable. It can be chosen differently; for example, θ can vary from 0 to 2π, but then the stream function will jump along the positive x-axis. The line along which the stream function jumps is called the branch cut. It appears that the contouring routine plots a thicker line along the branch cut. This is because all the stream function values that are contoured fall between the values computed at the grid points just above the negative x-axis and the values computed at the grid points just below the negative x-axis. The thick line is less pronounced when more grid points are used.

A plot with both equipotentials and streamlines is called a flow net (Figure 6.6). Equipotentials and streamlines always cross at right angles, unless the aquifer is anisotropic. It is common to use the same increments $\Delta\Phi = \Delta\Psi$ when constructing a flow net. Note that in the code below, the angle θ is computed with the np.arctan2 function, which computes the arctan function given a value of y and x (in that order) and returns the corresponding angle in the range $-\pi$ to $+\pi$. Before creating a contour plot, a subplot is created where the scale is set equal in the x- and y-directions using the aspect keyword so that, e.g., a circle indeed looks like a circle and not an ellipse.

```
# parameters
xw = 0 # x-location of well, m
yw = 0 # y-location of well, m
Q = 100 # discharge of well, m^3/d
```

```
# solution
xg, yg = np.meshgrid(np.linspace(-100, 100, 100), np.linspace(-100, 100, 100))
phi = Q / (4 * np.pi) * np.log(((xg - xw) ** 2 + (yg - yw) ** 2) / 100 ** 2)
psi = Q / (2 * np.pi) * np.arctan2(yg - yw, xg - xw)
```

```
# basic flow net (right graph)
plt.subplot(111, aspect=1)
plt.contour(xg, yg, psi, np.arange(-Q / 2, Q / 2, Q / 8))
plt.contour(xg, yg, phi, np.arange(phi.min(), phi.max(), Q / 8));
```

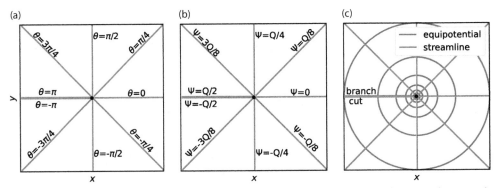

Figure 6.6   Angle θ measured in a local coordinate system centered at the well (left), stream function along eight streamlines including branch cut (middle), and flow net (right) for radial flow to a well.

As an example of the superposition principle, consider the case of an extraction well and an injection well, also referred to as a well doublet. Well 0 is located at $(x_0, y_0) = (-d, 0)$ and extracts with discharge $Q_0 = Q$. Well 1 is located at $(x_1, y_1) = (d, 0)$ and has an injection rate that is equal in magnitude to the discharge of well 0, which means that $Q_1 = -Q$. The head at $(x, y) = (0, 0)$ is equal to h_c, which is converted to a discharge potential Φ_c. The solution is obtained by the addition of two solutions (6.37) as (recall that $\ln(r_0^2) - \ln(r_1^2) = \ln(r_0^2/r_1^2)$)

$$\Phi = \frac{Q}{4\pi} \ln(r_0^2/r_1^2) + \Phi_c \tag{6.43}$$

where r_0 and r_1 are the distances to well 0 and well 1, respectively. Note that at $(x, y) = (0, 0)$, $r_0 = r_1$ so that the potential is indeed equal to Φ_c. Expressions for the discharge vector and the stream function are obtained in a similar fashion through superposition. A flow net and a streamplot are shown below. All water that is injected by the right well of the doublet flows to the left well of the doublet although the path may be very long. The branch cut in the stream function only extends from the injection well to the extraction well (left graph of Figure 6.7). To the left of the pumping well ($x \leq -d$), the branch cut of the pumping well cancels the branch cut of the injection well.

```
# parameters
T = 100 # transmissivity, m^2/d
hc = 0 # head at (0, 0)
d = 100 # half distance between the two wells, m
Q = [100, -100] # discharges of wells, m^3/d
```

```
xw = [-d, d] # x-locations of wells, m
yw = [0, 0] # y-locations of wells, m
phic = T * hc # discharge potential at origin, m^3/d
```

```
# solution
xg, yg = np.meshgrid(np.linspace(-200, 200, 100), np.linspace(-150, 150, 100))
pot = phic
psi = 0.0
Qx = 0.0
Qy = 0.0
for n in range(2):
    rsq = (xg - xw[n]) ** 2 + (yg - yw[n]) ** 2
    pot += Q[n] / (4 * np.pi) * np.log(rsq)
    psi += Q[n] / (2 * np.pi) * np.arctan2(yg - yw[n], xg - xw[n])
    Qx += -Q[n] / (2 * np.pi) * (xg - xw[n]) / rsq
    Qy += -Q[n] / (2 * np.pi) * (yg - yw[n]) / rsq
```

```
# basic plot
plt.figure(figsize=(8, 4))
plt.subplot(121, aspect=1)
plt.contour(xg, yg, pot, np.arange(phic - 100, phic + 100, 10))
plt.contour(xg, yg, psi, np.arange(-Q[0], Q[0], 10))
plt.subplot(122, aspect=1)
plt.streamplot(xg, yg, Qx, Qy);
```

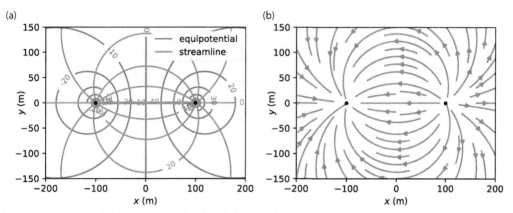

Figure 6.7 Flow net (left) and streamplot (right) for a well doublet.

Interestingly, the potential is constant and equal to Φ_c along the entire line $x = 0$. This makes sense by looking at Eq. (6.43). After all, everywhere along the line $x = 0$, the distance to well 0 is equal to the distance to well 1 so that $\ln(r_0^2/r_1^2) = 0$. This means that the left half of the flow field shown in Figure 6.7 ($x \leq 0$) represents the solution for a well pumping near a long and straight river with $\Phi = \Phi_c$ along $x = 0$. This method of obtaining the solution for a well near a river with constant head is known as the method of images.

Exercise 6.5: A well is located 65 m from a river that is roughly straight near the well. The well is installed in a 0.8-m-diameter hole. The well screen and the river are in direct contact with the same aquifer, which has an estimated transmissivity $T = 150$ m²/day. The well has been pumping for a long time at a rate of $Q = 600$ m³/day. Estimate the head in the well using the river level as the datum.

Exercise 6.6: For the example above, compute the stream function at $(x, y) = (0, 100)$, $(x, y) = (0, \varepsilon)$, $(x, y) = (0, -\varepsilon)$, and $(x, y) = (0, -100)$, where ε is a small increment, by determining the appropriate angles, and compare the answer to results with the `arctan2` function.

The method of images is an application of the superposition principle. Wells (or other singularities) are placed outside of the problem domain, and symmetry is used to satisfy conditions specified along boundaries. A well near a river with constant head is simulated by adding an image well on the other side of the river. The discharge of the image well must be equal, but opposite in sign to the discharge of the well, and the image well must be located at the "image location" of the well, that is, the same distance from the river, but on the other side. The solution for multiple wells near a river may be obtained in the same fashion, as long as a corresponding image well is added for each well.

A well near an impermeable rock formation (an impermeable or no-flow boundary) may be simulated by adding an image well with the same discharge on the other side of the impermeable boundary. An example of a well near an impermeable boundary along $x = 0$ is shown in Figure 6.8. The code for the solution and basic plot is the same as the code above, and is not repeated. Only the values of the discharge Q are changed.

```
# changed parameters
Q = [100, 100] # discharge of well and image well, m^3/d
```

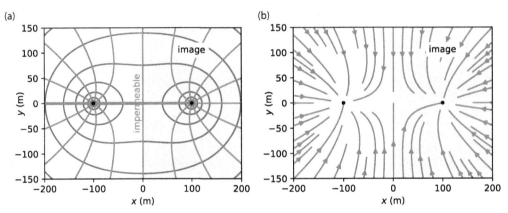

Figure 6.8 Flow net (left) and streamplot (right) for a well near an impermeable boundary.

As a final example of the method of images, consider flow to a well near a river with a head-specified boundary (constant discharge potential $\Phi = \Phi_c$) along the negative x-axis and an impermeable boundary along the negative y-axis (Figure 6.9). The well has discharge Q and is located at (x_0, y_0), where both x_0 and y_0 are negative. The solution for this problem requires three image wells. First, an image well with discharge Q is placed at $(-x_0, y_0)$, which ensures that the negative y-axis is an impermeable boundary. Next, image wells with discharge $-Q$ are placed at $(x_0, -y_0)$ and $(-x_0, -y_0)$ so that the negative x-axis is a specified head boundary, while the negative y-axis remains an impermeable boundary. The location and discharge of the well and three image wells are shown in the left side of Figure 6.9. The discharge potential may be written as

$$\Phi = \frac{Q}{2\pi} \ln \left(\frac{r_0 r_1}{r_2 r_3} \right) + \Phi_c \tag{6.44}$$

It is easy to verify that everywhere along the river (specified head) $r_0 = r_2$ and $r_1 = r_3$ so that, indeed, the potential (and hence the head) is constant. The boundary condition along the impermeable boundary may be verified by writing out an equation for Q_x. A flow net for the flow domain is created below. The code below creates a flow net with a pretty ugly branch cut along the line $y = -50$. This branch cut is not shown in Figure 6.9 by separately contouring the domain from $y = -150$ to just below $y = -50$ and then, on the same graph, from just above $y = -50$ to $y = 0$ (code is not shown).

```
# parameters
xw = [-100, 100, -100, 100] # x-locations of well and image wells, m
yw = [-50, -50, 50, 50] # y-locations of well and image wells, m
Qw = [100, 100, -100, -100] # discharge of well and image wells, m^3/d
```

```
# solution
xg, yg = np.meshgrid(np.linspace(-200, 0, 101), np.linspace(-150, 0, 101))
phi = 0
psi = 0
for n in range(4):
    phi += Qw[n] / (4 * np.pi) * np.log((xg - xw[n]) ** 2 + (yg - yw[n]) ** 2)
    psi += Qw[n] / (2 * np.pi) * np.arctan2(yg - yw[n], xg - xw[n])
```

```
# flow net in right graph
plt.subplot(111, aspect=1)
plt.contour(xg, yg, phi, np.arange(-100, 100, 5), colors='C0')
plt.contour(xg, yg, psi, np.arange(-200, 200, 5), colors='C1');
```

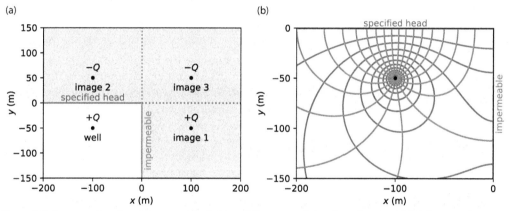

Figure 6.9 Layout of well and image wells (left) and flow net of the flow domain (right) for a well near a specified-head boundary and an impermeable boundary.

Many analytical solutions for problems with wells and equipotential boundaries and/or impermeable boundaries may be obtained by the method of images. The method of images works for many types of flow, including confined flow, unconfined flow, interface flow, semi-confined flow, and transient flow, and is also applicable to some problems with a Robin boundary condition or an inhomogeneity boundary.

Exercise 6.7: A doublet of a well and an injection well are located near an impermeable boundary located along the y-axis; the radius of both wells is 0.3 m. The extraction well is located at $(x, y) = (100, 0)$ and has a discharge $Q = 100$ m^3/d, and the injection well is located at $(x, y) = (50, 50)$ and has an injection rate 100 m^3/d. The aquifer is confined with transmissivity $T = 80$ m^2/d. Far away from the doublet, the head in the aquifer is zero. Draw a flow net, and compute the heads at the wells.

Exercise 6.8: A well is located a distance $d = 50$ m from a long and straight river with water level $h_r = 20$ m. The river is in direct contact with the aquifer, and the radius of the well is 0.3 m. The transmissivity of the aquifer may be approximated as constant and equal to

$T = 200$ m^2/d. Compute the discharge of the well such that the head at the well is 2 m below the water level in the river.

Exercise 6.9: Consider a well pumping with discharge Q on a circular island with radius R; the radius of the well is r_w. The origin of the coordinate system is at the center of the island, and the well is located at $(x, y) = (d, 0)$. The head at the edge of the island is h_0. The transmissivity of the aquifer may be approximated as constant and equal to T. An expression for the head on the island may be obtained with the method of images for a circle (Strack, 1989)

$$h = \frac{Q}{4\pi T} \ln \left[\frac{(x-d)^2 + y^2}{(x - R^2/d)^2 + y^2} \frac{R^2}{d^2} \right] + h_0 \tag{6.45}$$

Compute and draw head contours, and compute the head at the well. Given: $T = 200$ m^2/d, $R = 500$, m, $d = 200$ m, $Q = 500$ m^3/d, $r_w = 0.3$ m, and $h_0 = 10$ m.

6.3 Wells near an inhomogeneity boundary

Hydraulic properties always vary at least a little bit throughout an aquifer. A common approach to representing varying aquifer properties is to divide the aquifer up into zones of different, but homogeneous properties; each zone is referred to as an inhomogeneity. Inside an inhomogeneity, the hydraulic conductivity, base elevation, and/or thickness of the aquifer may be different from its surroundings. As such, the transmissivity of the aquifer is referred to as piecewise constant.

The boundary conditions at the inhomogeneity boundary are that both the head and the normal component of flow are continuous across it. As a result, the component of flow parallel to the inhomogeneity boundary jumps. This is a direct consequence of the condition that the head is continuous. For example, consider an inhomogeneity boundary along the y-axis, as shown in Figure 6.10. The transmissivity to the left of the boundary is T_L, and the transmissivity to the right of the boundary is T_R. The head is the same on either side of the inhomogeneity boundary, which means that also the head gradient $\partial h/\partial y$ is continuous across the boundary. As a result, the component of the discharge vector parallel to the boundary, in this case Q_y, is $Q_y^- = -T_L \partial h/\partial y$ on the left side of the boundary and $Q_y^+ = -T_R \partial h/\partial y$ on the right side of the boundary. The jump in the component of flow parallel to the inhomogeneity boundary is referred to as a shear flow.

The direction of a streamline changes abruptly when it crosses an inhomogeneity boundary. This is a direct consequence of the shear flow along the boundary. For example, consider one point on the inhomogeneity boundary along the y-axis of Figure 6.10. The flow just to the left of the boundary is (Q_x^-, Q_y^-), and the flow just to the right of the boundary is (Q_x^+, Q_y^+). The angle θ^- between $\vec{Q}^-$ and the normal to the boundary is

$$\tan(\theta^-) = Q_y^- / Q_x^- \tag{6.46}$$

Similarly, the angle θ^+ between $\vec{Q}^+$ and the normal to the boundary is

$$\tan(\theta^+) = Q_y^+ / Q_x^+ \tag{6.47}$$

Since $Q_x^- = Q_x^+$, this gives

$$\frac{\tan(\theta^-)}{\tan(\theta^+)} = \frac{T_L}{T_R} \tag{6.48}$$

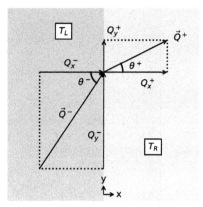

Figure 6.10 Refraction of the discharge vector at an inhomogeneity boundary for the case that $T_R < T_L$.

This means that when water flows from higher transmissivity to lower transmissivity ($T_R < T_L$, as shown in Figure 6.10), the angle between the discharge vector and the normal to the boundary decreases. It is said that the discharge vector is refracted toward the normal.

As an example, consider a well pumping near an inhomogeneity boundary along the y-axis, as discussed above. The well is located at $(x, y) = (-d, 0)$ and is numbered well 0. The solution for an inhomogeneity boundary is obtained with the method of images by placing an image well at $(x, y) = (d, 0)$ and choosing the discharge of the image well such that the boundary conditions along the inhomogeneity boundary are met. The solution for the head is (e.g., Strack, 2017, Section 8.10)

$$h = \frac{Q}{2\pi T_L} \ln\left(\frac{r_0}{r_1}\right) + \frac{Q}{\pi(T_L + T_R)} \ln(r_1) \qquad x \le 0 \tag{6.49}$$

$$h = \frac{Q}{\pi(T_L + T_R)} \ln(r_0) \qquad x \ge 0 \tag{6.50}$$

where $r_0 = \sqrt{(x + d)^2 + y^2}$ and $r_1 = \sqrt{(x - d)^2 + y^2}$. The solution for the stream function is

$$\Psi = \frac{Q}{2\pi}(\theta_0 - \theta_1) + \frac{QT_L}{\pi(T_L + T_R)}\theta_1 \qquad x \le 0 \tag{6.51}$$

$$\Psi = \frac{QT_R}{\pi(T_L + T_R)}\theta_0 \qquad x \ge 0 \tag{6.52}$$

where $\theta_0 = \arctan(y/(x + d))$ and $\theta_1 = \arctan(-y/(d - x))$. Note that θ_0 varies from $-\pi$ to π, while θ_1 varies from 0 to 2π.

An example is shown for both the case that $T_R = T_L/5$ and the case that $T_R = 5T_L$ (Figure 6.11). The total discharge that the well pumps from the zone on the right, where $T = T_R$, is $T_R Q/(T_L + T_R)$. For the case that T_R is small relative to T_L, the inhomogeneity boundary starts to represent an impermeable boundary. For the case that T_R is large relative to T_L, the inhomogeneity boundary starts to represent an equipotential. Curiously, for this case of a well pumping near a straight inhomogeneity boundary, the streamlines on the right side of the inhomogeneity boundary are straight and directed toward the well.

```
# parameters
d = 100 # distance of well from inhomogeneity boundary, m
Q = 100 # discharge of well, m^3/d
TL = 100 # transmissivity on left side, m^2/d
```

```
# solution
def head_and_psi(x, y, TL, TR, d=100, Q=100):
    r0 = np.sqrt((x + d) ** 2 + y ** 2)
    r1 = np.sqrt((x - d) ** 2 + y ** 2)
    theta0 = np.arctan2(y, x + d)
    theta1 = np.arctan2(-y, d - x)
    if x < 0:
        h = Q / (2 * np.pi * TL) * np.log(r0 / r1) + \
            Q / (np.pi * (TL + TR)) * np.log(r1)
        psi = Q / (2 * np.pi) * (theta0 - theta1) + \
            Q * TL / (np.pi * (TL + TR)) * theta1
    else:
        h = Q / (np.pi * (TL + TR)) * np.log(r0)
        psi = Q * TR / (np.pi * (TL + TR)) * theta0
    return h, psi

hpsifunc = np.vectorize(head_and_psi)

xg, yg = np.meshgrid(np.linspace(-200, 200, 100), np.linspace(-150, 150, 100))
ha, psia = hpsifunc(xg, yg, TL=TL, TR=TL / 5, d=d, Q=Q)
hb, psib = hpsifunc(xg, yg, TL=TL, TR=5 * TL, d=d, Q=Q)
```

```
# basic plot
plt.subplot(121, aspect=1, title='$T_L$=100 m$^2$/d, $T_R$=20 m$^2$/d')
plt.contour(xg, yg, ha, 40, colors='C0')
plt.contour(xg, yg, psia, np.arange(-50, 50, 2.5), colors='C1')
plt.fill([0, 200, 200, 0], [-150, -150, 150, 150], fc=[.9, .9, .9])
plt.subplot(122, aspect=1, title='$T_L$=100 m$^2$/d, $T_R$=500 m$^2$/d')
plt.contour(xg, yg, hb, 40, colors='C0')
plt.contour(xg, yg, psib, np.arange(-50, 50, 2.5), colors='C1')
plt.fill([0, 200, 200, 0], [-150, -150, 150, 150], fc=[.9, .9, .9]);
```

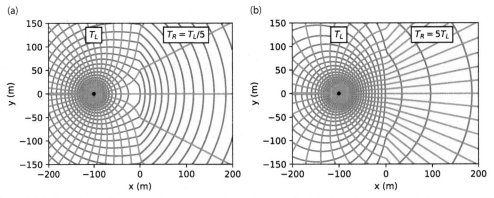

Figure 6.11   Pumping well near an inhomogeneity boundary; on the left side, the gray zone is less permeable, while on the right side, the gray zone is more permeable; contour interval of the stream function is $Q/40$.

6.4   Wells in a semi-confined aquifer

Steady flow to a well in the bottom aquifer of a two-aquifer system is considered in the next two sections. The two aquifers are separated by a leaky layer. In this section, the head in the top aquifer is fixed, for example by a system of ditches and drains, so that the bottom aquifer is a semi-confined aquifer (Figure 6.12). The case for a two-aquifer system, where water is flowing through both aquifers, is presented in the next section. Recall that solutions for semi-confined flow and two-aquifer flow were written in terms of heads in Chapter 2 as the discharge potentials defined in Table (6.1) cannot be used.

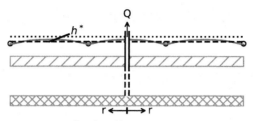

Figure 6.12 A well pumping in a semi-confined aquifer.

The differential equation for steady radially symmetric semi-confined flow is

$$\nabla^2 h = \frac{\mathrm{d}^2 h}{\mathrm{d}r^2} + \frac{1}{r}\frac{\mathrm{d}h}{\mathrm{d}r} = \frac{h - h^*}{cT} \tag{6.53}$$

where T is the transmissivity of the aquifer, h^* is the constant head above the semi-confining layer, and c is the resistance to vertical flow of the semi-confining layer (see Chapter 2). A solution is sought for radial flow to a fully penetrating well with discharge Q. The boundary condition at the well is the same as that for a well in confined flow

$$\lim_{r \to 0} -2\pi r Q_r = Q \tag{6.54}$$

The boundary condition at infinity is that the head is equal to the fixed head h^* in the top aquifer

$$h|_{r=\infty} = h^* \tag{6.55}$$

The solution is (De Glee, 1930)

$$h = h^* - \frac{Q}{2\pi T}\mathrm{K}_0(r/\lambda) \tag{6.56}$$

where the leakage factor is $\lambda = \sqrt{cT}$ (see Chapter 2) and K_0 is the modified Bessel function of the second kind and order 0. The Bessel function $\mathrm{K}_0(r/\lambda)$ approaches zero for r approaching infinity. In practice, K_0 can be approximated as zero for $r > 4\lambda$, as will be shown later.

The main difference between the solutions for pumping in a confined aquifer and pumping in a semi-confined aquifer is that in a confined aquifer, all water pumped by the well comes from infinity (and hence the head keeps increasing toward infinity), while in a semi-confined aquifer, all water pumped by the well comes from the overlying aquifer by leakage through the semi-confining layer.

The radial component of the discharge vector is obtained from differentiation

$$Q_r = -T\frac{\mathrm{d}h}{\mathrm{d}r} = \frac{-Q}{2\pi\lambda}\mathrm{K}_1(r/\lambda) \tag{6.57}$$

where it is used that $\mathrm{d}\mathrm{K}_0/\mathrm{d}r = -\mathrm{K}_1$, where K_1 is the modified Bessel function of the second kind and order 1. It can be shown that Q_r fulfills boundary condition (6.54) by using that K_1 may be approximated near $r = 0$ as (e.g., Abramowitz and Stegun, 1965)

$$\mathrm{K}_1 \approx \lambda/r \qquad r/\lambda \ll 0 \tag{6.58}$$

The total flow Q_{tot} in the aquifer toward the well as a function of the radial distance r may be computed as $-Q_r$ multiplied by the circumference of a circle with radius r

$$Q_{tot} = -2\pi r Q_r \tag{6.59}$$

Substitution of Eq. (6.57) for Q_r using Eq. (6.58) shows that boundary condition (Eq. 6.54) is indeed fulfilled

$$Q_{tot} = -2\pi r \frac{-Q}{2\pi \lambda} \frac{\lambda}{r} = Q \qquad r/\lambda \ll 0 \tag{6.60}$$

The head near the well may be approximated as

$$h = h^* + \frac{Q}{2\pi T} \ln\left(\frac{r}{1.123\lambda}\right) \qquad (r/\lambda \ll 1) \tag{6.61}$$

where it is used that (Abramowitz and Stegun, 1965)

$$K_0(r/\lambda) \approx -\ln\left(\frac{re^\gamma}{2\lambda}\right) \approx -\ln\left(\frac{r}{1.123\lambda}\right) \qquad (r/\lambda \ll 1) \tag{6.62}$$

where $\gamma = 0.5772...$ is Euler's constant. Note that Eq. (6.61) is the Thiem solution (6.25) where the head is equal to h^* at a distance of $R = 1.123\lambda$; the approximate formula gives accurate results near the well. The Bessel functions K_0 and K_1 are available as k0 and k1 in the scipy.special package.

The solution for a well (Eq. 6.56) and the approximate formula (6.61) near the well are plotted in the left graph of Figure 6.13. In the right graph, Q_{tot} is plotted as a function of r/λ. Note that Q_{tot}/Q is indeed equal to 1 at the well and reduces to $Q_{tot}/Q = 0.05$ at $r = 4\lambda$. For practical purposes, the effect of the well may be neglected beyond a distance of 4λ.

```
# parameters
T = 200 # transmissivity of bottom aquifer, m^2/d
c = 1000 # resistance of leaky layer, d
hstar = 0 # fixed head in top aquifer, m
rw = 0.3 # radius of the well, m
lab = np.sqrt(T * c) # leakage factor, m
Q = 500 # discharge of well, m^3/d
```

```
# solution
from scipy.special import k0, k1
r = np.linspace(rw, 4 * lab, 200)
h = -Q / (2 * np.pi * T) * k0(r / lab) + hstar
happrox = Q / (2 * np.pi * T) * np.log(r / (1.123 * lab)) + hstar
Qtot = 2 * np.pi * r * Q / (2 * np.pi * lab) * k1(r / lab)
```

```
# basic plot
plt.subplot(121)
plt.plot(r / lab, h, label='head')
plt.plot(r / lab, happrox, label='approx')
plt.legend()
plt.subplot(122)
plt.plot(r / lab, Qtot / Q);
```

Exercise 6.10: For the values of the example, compute the head at the well as a function of c for c varying from 100 d to 5000 d. Plot the head at the well vs. c.

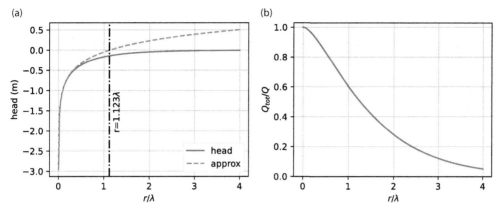

Figure 6.13 A well in a semi-confined aquifer; head (left) and total flow in the aquifer (right).

Exercise 6.11: Consider two pumping wells in a semi-confined aquifer. Aquifer properties are $T = 200$ m²/d, $c = 200$ d, $h^* = -2$ m. Well 0 is located at $(x_0, y_0) = (0, 0)$ and has a discharge $Q_0 = 500$ m³/d, and well 1 is located at $(x_1, y_1) = (400, 200)$ and has a discharge $Q_1 = 200$ m³/d; the radius of the wells is 0.2 m. Make a contour plot of the heads in the aquifer. Compute the head at wells 0 and 1 for the case that $c = 200$ d. Next, compute the head at wells 0 and 1 for the case that $c = 1000$ d.

6.5   Wells in a two-aquifer system

In the previous section, the problem was solved for a well that pumps from the bottom aquifer of a two-aquifer system where the head in the top aquifer is constant and fixed. Here, a solution is presented for flow to a well in a two-aquifer systems where water flows in both aquifers. The well is screened in the bottom aquifer. The aquifers are numbered 0 and 1 from the top (see Figure 6.14). The transmissivities of the top and bottom aquifers are T_0 and T_1, respectively, and the resistance of the leaky layer between the aquifers is c.

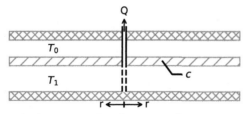

Figure 6.14 A well pumping in the bottom aquifer of a two-aquifer system.

The flow in the two-aquifer system is governed by a system of two coupled differential equations (see Section 2.4)

$$\nabla^2 h_0 = \frac{h_0 - h_1}{cT_0}$$
$$\nabla^2 h_1 = \frac{h_1 - h_0}{cT_1} \tag{6.63}$$

where h_0 and h_1 are the heads in the top and bottom aquifers, respectively. A well is pumping with a discharge Q in aquifer 1

$$\lim_{r \to 0} -2\pi r Q_{r,1} = Q \tag{6.64}$$

The second boundary condition is that the head is fixed to h_R at $r = R$ in aquifer 0

$$h_0|_{r=R} = h_R \tag{6.65}$$

The solution for the head in each aquifer is a combination of the solution for a well in a confined aquifer (Eq. 6.37) and the solution for a well in a semi-confined aquifer (Eq. 6.56)

$$h_0 = \frac{Q}{2\pi(T_0 + T_1)}[\ln(r/R) + K_0(r/\lambda) - K_0(R/\lambda)] + h_R$$
$$h_1 = \frac{Q}{2\pi(T_0 + T_1)}\left[\ln(r/R) - \frac{T_0 K_0(r/\lambda)}{T_1} - K_0(R/\lambda)\right] + h_R \tag{6.66}$$

where

$$\lambda = \sqrt{\frac{c T_0 T_1}{T_0 + T_1}} \tag{6.67}$$

The corresponding radial components of the discharge vector are

$$Q_{r,0} = -\frac{T_0 Q}{2\pi(T_0 + T_1)}\left[\frac{1}{r} - \frac{K_1(r/\lambda)}{\lambda}\right]$$
$$Q_{r,1} = -\frac{T_1 Q}{2\pi(T_0 + T_1)}\left[\frac{1}{r} + \frac{T_0 K_1(r/\lambda)}{T_1 \lambda}\right] \tag{6.68}$$

Near the well, all flow is in aquifer 1, as can be seen from taking the limit and using that $K_1(r/\lambda) \approx \lambda/r$ for small r

$$\lim_{r\to 0} 2\pi r Q_{r,0} = 0$$
$$\lim_{r\to 0} 2\pi r Q_{r,1} = Q \tag{6.69}$$

Far away from the well (in practice, beyond 4λ), the flow is distributed between the two aquifers according to their transmissivities

$$\lim_{r\to\infty} 2\pi r Q_{r,0} = -\frac{T_0}{T_0 + T_1}Q$$
$$\lim_{r\to\infty} 2\pi r Q_{r,1} = -\frac{T_1}{T_0 + T_1}Q \tag{6.70}$$

In the example below, the head and total flow are plotted as a function of the radial distance (Figure 6.15). The head in the bottom aquifer shows a sharp drawdown at the well ($r = 0$), while the head gradient equals zero at the well location in the top aquifer. The graph of the flow represents the total flow $Q_{\text{tot}} = -2\pi r Q_r$ in the aquifer as a function of the radial distance. The sum of the total flow in the two aquifers always equals Q. At the well, all flow is in the bottom aquifer, while far away from the well, the flow is distributed according to the transmissivities, as explained above. A streamline plot is shown for a cross section through the well in Figure 6.16.

```
# parameters
T0 = 100 # transmissivity aquifer 0, m^2/d
T1 = 150 # transmissivity aquifer 1, m^2/d
c  = 1000 # resistance of leaky layer, d
```

```
zaq = [0, -10, -15, -25] # elevation of top of layers and bottom, m
R = 200 # distance where head is fixed, m
hR = 5 # head at r=R, m
Q = 1000 # discharge of well, m^3/d
rw = 0.3 # radius of well, m
lab = np.sqrt(c * T0 * T1 / (T0 + T1)) # leakage factor, m
```

```
# solution
from scipy.special import k0, k1
r = np.linspace(rw, 5 * lab, 200)
h0 = Q / (2 * np.pi * (T0 + T1)) * (np.log(r / R) + \
     k0(r / lab) - k0(R / lab)) + hR
h1 = Q / (2 * np.pi * (T0 + T1)) * (np.log(r / R) + \
     T0 * k0(r / lab) / T1 - k0(R / lab)) + hR
Q0tot = -Q * T0 / (2 * np.pi * (T0 + T1)) * (1 / r -
        k1(r / lab) / lab) * 2 * np.pi * r
Q1tot = -Q * T1 / (2 * np.pi * (T0 + T1)) * (1 / r +
        T0 * k1(r / lab) / (T1 * lab)) * 2 * np.pi * r
psi = np.zeros((len(zaq), len(r)))
psi[1] = Q0tot
psi[2] = Q0tot
psi[3] = Q1tot + Q0tot
```

```
# basic plot head and flow
plt.subplot(121)
plt.plot(r / lab, h0, label='aquifer 0')
plt.plot(r / lab, h1, label='aquifer 1')
plt.legend()
plt.subplot(122)
plt.plot(r / lab, Q0tot / Q, label='aquifer 0')
plt.plot(r / lab, Q1tot / Q, label='aquifer 1')
plt.legend();
```

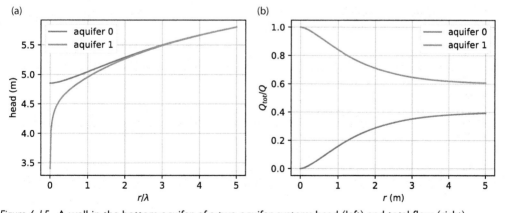

Figure 6.15 A well in the bottom aquifer of a two-aquifer system; head (left) and total flow (right).

```
# basic streamline plot for r >= 0
plt.contour(r, zaq, psi, 10);
```

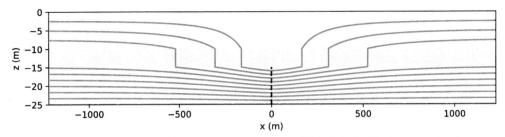

Figure 6.16 Streamlines in a vertical cross section through a well (black dashed line) screened in the bottom aquifer of a two-aquifer system.

Exercise 6.12: Consider an extraction well and an injection well pumping in the bottom aquifer. All parameters are as in the example. The extraction well is located at $(x, y) = (-100, 0)$ and has a discharge $Q = 1000$ m^3/d, and the injection well is located at $(x, y) = (100, 0)$ and has a discharge $Q = -1000$ m^3/d; the radius of both wells is 0.3 m. The head far away from the doublet is 5 m. Plot the head along the line $y = 0$, and compute the head at the extraction well in both aquifers.

Chapter 7

Steady two-dimensional flow to wells in uniform background flow

Groundwater always moves in some direction inside an aquifer (although sometimes imperceptibly slow) prior to the installation of a well. This is referred to as the background flow. In the previous chapter, wells were pumping in aquifers where there was no flow in the absence of the wells. In this chapter, solutions are presented for the case that the background flow is uniform. A number of cases of wells in an otherwise uniform flow are considered, including wells near rivers and streams, and wells near the coast. The latter problems are solved through the use of a complex potential, which is a convenient and efficient way to combine the discharge potential and stream function into one complex function. Special attention is paid to the location of stagnation points (points of zero flow) in the flow field.

7.1 A single well in uniform background flow

The discharge potential for uniform flow at a rate U in the x-direction and V in the y-direction is

$$\Phi = -Ux - Vy \tag{7.1}$$

so that indeed

$$Q_x = -\frac{\partial \Phi}{\partial x} = U \qquad Q_y = -\frac{\partial \Phi}{\partial y} = V \tag{7.2}$$

The corresponding stream function is

$$\Psi = Vx - Uy \tag{7.3}$$

which also gives

$$Q_x = -\frac{\partial \Psi}{\partial y} = U \qquad Q_y = \frac{\partial \Psi}{\partial x} = V \tag{7.4}$$

Exercise 7.1: Consider uniform flow in a confined aquifer with $T = 200$ m²/d. The head is measured at three locations: $h(0, 0) = 20$ m, $h(100, 0) = 19.8$ m, and $h(0, 100) = 19.9$ m. Compute the x- and y-components of the uniform flow, and make a flow net.

The potential and stream function for a well pumping with a discharge Q located at (x_w, y_w) in an otherwise uniform flow field U in the x-direction (i.e., $Q_y = 0$) may be obtained by superposition of the solution for uniform flow (Eqs. 7.1 and 7.3) and the solution for a well

DOI: 10.1201/9781315206134-7

(Eqs. 6.37 and 6.41)

$$\Phi = -Ux + \frac{Q}{2\pi}\ln(r) + C \tag{7.5}$$

$$\Psi = -Uy + \frac{Q}{2\pi}\theta \tag{7.6}$$

where $r = \sqrt{(x-x_w)^2 + (y-y_w)^2}$, $\theta = \arctan[(y-y_w)/(x-x_w)]$, and C is a constant to be determined from additional boundary conditions. The corresponding components of the discharge vector are (independent of the value of C)

$$Q_x = -\frac{\partial \Phi}{\partial x} = U - \frac{Q}{2\pi}\frac{x-x_w}{r^2} \tag{7.7}$$

$$Q_y = -\frac{\partial \Phi}{\partial y} = -\frac{Q}{2\pi}\frac{y-y_w}{r^2} \tag{7.8}$$

The well captures part of the background flow. The capture zone envelope is the dividing streamline that bounds the area of the aquifer that is (eventually) captured. The capture zone envelope goes through the point downstream of the well where $Q_x = Q_y = 0$. This point is referred to as a stagnation point. The location of the stagnation point (x_s, y_s) is found by setting $Q_x = 0$ along the line $y = y_w$ ($Q_y = 0$ along the entire line $y = y_w$)

$$Q_x|_{x=x_s,y=y_w} = U - \frac{Q}{2\pi}\frac{1}{(x_s - x_w)} = 0 \tag{7.9}$$

where it is used that $r^2 = (x_s - x_w)^2$ along the line $y = y_w$. The x-location of the stagnation point is obtained as

$$x_s = x_w + \frac{Q}{2\pi U} \tag{7.10}$$

The flow field for a well in uniform background flow is shown in the left graph of Figure 7.1. The stagnation point is at the green cross. For this case, $\Psi = -Uy_w$ at the stagnation point, where $\theta = 0$ and $y = y_w$. The streamline representing the capture zone envelope (shown in black) goes through the stagnation point so that $\Psi = -Uy_w$ along the entire capture zone envelope for this case. The width w of the capture zone far upstream of the well may be obtained from continuity of flow, as the amount of flow between the two dividing streamlines must equal the discharge of the well, $Uw = Q$, so that

$$w = Q/U \tag{7.11}$$

A close-up of the flow around the stagnation point is shown in the right graph of Figure 7.1. Two parts of the streamline representing the capture zone envelope are directed toward the stagnation point, while the other two parts are directed away from the stagnation point. One of the latter two parts flows to the well, and the other part flows away from the well.

```
# parameters
xw = 0 # x-location of well, m
yw = 0 # y-location of well, m
Q = 80 # discharge of well, m^3/d
U = 0.5 # uniform flow in x-direction, m^2/d
```

```
# solution
xs = xw + Q / (2 * np.pi * U)
xg, yg = np.meshgrid(np.linspace(-200, 100, 100), np.linspace(-100, 100, 100))
psi = Q / (2 * np.pi) * np.arctan2(yg, xg) - U * yg
```

```
# basic streamline plot, left graph
plt.subplot(111, aspect=1)
plt.contour(xg, yg, psi, 10, colors='C1', linestyles='-')
plt.contour(xg, yg, psi, [-U * yw], colors='k', linestyles='-') # envelope
plt.plot(xs, 0, 'C2x'); # stagnation point
```

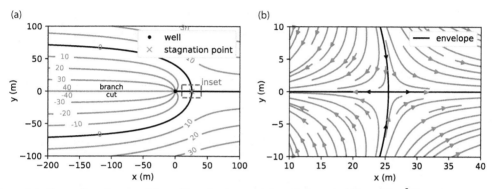

Figure 7.1   Streamlines for a well in uniform background flow (left, with $\Delta \Psi = 10$ m^3/d) and close-up of the inset around the stagnation point (right); the capture zone envelope is shown in black.

Exercise 7.2: Consider a well located in a field of uniform background flow. The uniform flow is directed in the positive x-direction, and the head drop is 1 m every 1000 m. The transmissivity of the aquifer is constant and equal to $T = 200$ m^2/d. A contaminated area is located near the well as shown in Figure 7.2. Compute the maximum discharge of the well such that the well does not capture any water that flows through the contaminated area. Plot the capture zone envelope to verify your answer.

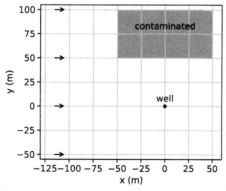

Figure 7.2   Setup for Exercise 7.2.

Exercise 7.3: Consider an extraction well and an injection well with equal, but opposite discharge $Q = 200$ m^3/d. The injection well is located at $(x, y) = (-50, 0)$, and the extraction well, at $(x, y) = (50, 0)$. The two wells are located in an otherwise uniform flow from west to east; the gradient far away is $\partial h/\partial x = -0.002$. The transmissivity of the aquifer is $T = 200$ m^2/d. Compute the locations of the two stagnation points, and draw the capture zone envelope.

Exercise 7.4: Consider two extraction wells with equal discharge $Q = 100$ m^3/d. One well is located at $(x, y) = (-50, 0)$, and the other well, at $(x, y) = (50, 0)$. The two wells are located in an otherwise uniform flow from west to east; the gradient far away is $\partial h/\partial x = -0.002$. The transmissivity of the aquifer is $T = 200$ m^2/d. Compute the locations of the two stagnation points, and draw the capture zone envelopes.

The average path of a water particle in the aquifer is referred to as a pathline. Unlike streamlines, pathlines can be computed for steady flow with and without areal infiltration and for transient flow. For steady two-dimensional flow without infiltration, pathlines and streamlimes coincide. For most cases of two-dimensional flow, pathlines cannot be computed analytically, but can be computed through numerical integration of the horizontal velocity vector $\vec{v} = (v_x, v_y)$

$$\frac{\mathrm{d}x}{\mathrm{d}t} = v_x$$
$$\frac{\mathrm{d}y}{\mathrm{d}t} = v_y \tag{7.12}$$

This is a system of two ordinary first-order differential equations, which can be integrated with a standard routine, for example the `solve_ivp` of `scipy.integrate`. Numerical integration results in an array of coordinates of the pathline including the travel time. Application of `solve_ivp` is discussed in some detail in the following.

The `solve_ivp` function requires three input arguments:

1   `fun(t, y)`, the function to be integrated, where for the case under consideration here, t is time and y is an array of x, y coordinates (not to be confused with the y-coordinates of the problem).
2   `t_span`, a two-tuple with the beginning and end times of the integration.
3   `y0`, an array with starting locations x, y.

In addition, `solve_ivp` has many keyword arguments to control, for example, the algorithm and accuracy of the numerical integration. Two keyword arguments are used here:

1   `t_eval`, which is an array with all times for which the integration is computed and returned. All values must be between the beginning and end times defined in `t_span`.
2   `events`, which is a list of functions to deal with pathlines that terminate before the end time of the integration is reached, for example when a pathline terminates at a well. Each function represents an event (e.g., the pathline has reached a well). Each event function takes t and y as input, just like the velocity function, and returns a continuous function that equals zero when the event occurs. For a well, the function returns the distance to the well minus the radius of the well so that it returns zero when the pathline is at the radius of the well. In addition, the function has an attribute `terminal` which is set to True so that the `solve_ivp` routine knows to terminate the integration when this event occurs (it is a bit unusual to specify an attribute for a function, but it is proper Python and quite useful here). If the `terminal` attribute is not set, the integration continues after the event occurs. This may be useful, e.g., when it is of interest when a pathline crosses a line.

The `solve_ivp` function returns an object with a number of attributes, including:

1   t, an array with the time points for which (x, y) values are returned.
2   y, a 2D array with the computed (x, y) coordinates of the pathline.

3   t_events, a list with arrays of times when events occurred.
4   y_events, a list with arrays of (x, y) values corresponding to the time values in
     t_events.
5   status, an integer -1 (failed), 0 (successfully reached end of tspan), or 1 (termination
     event occurred).
6   message, a readable message for termination.

In the example below, pathlines are computed for a well in uniform flow. The aquifer is confined so that the velocity vector is obtained as $\vec{v} = \vec{Q}/(nH)$. The velocity function is written,
and one event function is written to determine if a pathline reaches the well. Two pathlines are
started. Each pathline is followed for 10 years, but one of the pathlines terminates at the well
before the 10 years is reached. Note that in case an event occurs, the last time and coordinate
that are stored in attributes t and y are the last time and coordinate before the event occurs.
The time and coordinates of the event are stored separately in the t_events and y_events
attributes.

```
# additional parameters
n = 0.3 # porosity, -
H = 20 # aquifer thickness, m
rw = 0.3 # radius of well, m
tmax = 10 * 365 # total travel time, d
t = np.arange(0, 10 * 365 + 1, 365 / 20) # times pathline is returned, d
```

```
# functions
def vxvy(t, xy, xw=0, yw=0):
    x, y = xy
    rsq = (x - xw)**2 + (y - yw)**2
    Qx = U - Q / (2 * np.pi) * (x - xw) / rsq
    Qy = -Q / (2 * np.pi) * (y - yw) / rsq
    return np.array([Qx, Qy]) / (n * H)

def reached_well(t, xy):
    return np.sqrt((xy[0] - xw) ** 2 + (xy[1] - yw) ** 2) - rw

reached_well.terminal = True # specify the terminal attribute
```

```
# pathline solution
from scipy.integrate import solve_ivp
path0 = solve_ivp(vxvy, (0, tmax), y0=[-200, 50], t_eval=t, events=reached_well)
print('path0 message:', path0.message)
print(f'path0 reached well after {path0.t_events[0][0] / 365: .2f} years')
print(f'path0 entered well screen at', path0.y_events)
path1 = solve_ivp(vxvy, (0, tmax), y0=[-200, 80], t_eval=t, events=reached_well)
print('path1 message:', path1.message)
```

```
path0 message: A termination event occurred.
path0 reached well after  5.57 years
path0 entered well screen at [array([[0.17610724, 0.24287083]])]
path1 message: The solver successfully reached the end of the integration
interval.
```

```
# basic plot
plt.subplot(111, aspect=1)
plt.plot(path0.y[0], path0.y[1], label='pathline 0')
```

```
plt.plot(path1.y[0], path1.y[1], label='pathline 1')
plt.legend();
```

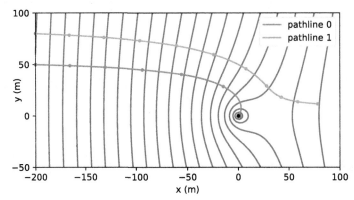

Figure 7.3 Pathlines with one-year travel time tick marks from the start at $x = -200$.

7.2 Well capture zones

All water inside the capture zone envelope of a well is eventually captured by the well, but this can take a (very) long time. In groundwater management, it is important to delineate the zone that is captured by the well within a certain time period. For example, all the water in the 5-year capture zone reaches the well within 5 years. The capture zone for a certain time period can be computed by integrating time along a streamline from the well against the flow. For a single well in an otherwise uniform flow field, this can be done analytically, as was shown by Bear and Jacobs (1965). For more complicated settings, the integration is conducted numerically. In this section, the analytical solution is discussed first, followed by the numerical solution.

Consider a well with discharge Q located at the origin in an otherwise uniform flow field with $Q_x = U$ and $Q_y = 0$. The stream function for this problem is given by Eq. (7.6). The stream function equals zero along the capture zone envelope (see previous section) so that an equation for the capture zone envelope can be written as

$$\Psi = -Uy + \frac{Q}{2\pi}\theta = 0 \tag{7.13}$$

Using that $y = r\sin\theta$ gives an expression for the radial distance from the well toward the capture zone envelope

$$r = \frac{Q}{2\pi}\frac{\theta}{U\sin(\theta)} \tag{7.14}$$

This means that for any angle θ, a distance r to the capture zone envelope can be computed with the above formula. The corresponding x, y coordinates of the capture zone envelope are computed with

$$x = r\cos(\theta) \qquad y = r\sin(\theta) \tag{7.15}$$

The travel time t from location (x, y) inside the capture zone envelope to the well can be computed with the following equation (Bear and Jacobs, 1965)

$$t = -\frac{nHx}{U} - \frac{QnH}{2\pi U^2}\ln\left[\frac{\sin(\theta - 2\pi Uy/Q)}{\sin(\theta)}\right] \tag{7.16}$$

where H is the aquifer thickness and n is the porosity. In the example below, the travel time toward the well is computed for a number of points inside the capture zone envelope, after which the capture zones are obtained by contouring the travel times (Figure 7.4).

```
# parameters
n = 0.25 # porosity, -
k = 20 # hydraulic conductivity, m/d
H = 15 # aquifer thickness, m
dhdx = -0.002 # gradient, -
U = -k * H * dhdx # uniform flow, m^2/d
Q = 500 # discharge of the well, m^3/d
rw = 0.3 # radius of well, m
```

```
# solution
# coordinates of capture zone envelope
theta = np.linspace(-3, -1e-4, 100) # bottom half of envelope
r = Q / (2 * np.pi) * theta / (U * np.sin(theta))
xcap = r * np.cos(theta)
ycap = r * np.sin(theta)
# grid inside the capture zone envelope
xg = xcap * np.ones((50, len(xcap)))
yg = np.zeros_like(xg)
for i in range(len(xcap)):
    yg[:, i] = np.linspace(ycap[i] + 1e-3, -ycap[i] - 1e-3, 50)
# travel time for all grid points inside capture zone envelope
theta = np.arctan2(yg, xg)
tgrid = -n * H * xg / U - Q * n * H / (2 * np.pi * U ** 2) * np.log(
        np.sin(theta - 2 * np.pi * U * yg / Q) / np.sin(theta))
```

```
# basic capture zone plot
plt.subplot(111, aspect=1, xlim=(-1000, 150))
plt.plot(0, 0, 'k.')
plt.contourf(xg, yg, tgrid, [0, 5 * 362.25, 10 * 365.25]);
```

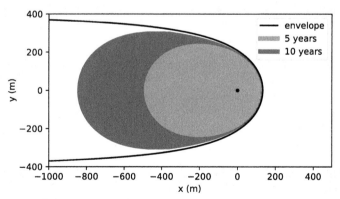

Figure 7.4 Capture zones for a well in a uniform background flow obtained with the analytic solution.

Capture zones for more complicated settings, e.g., multiple wells or other boundary conditions, must be computed numerically as they cannot be computed analytically. As was shown in the previous section, pathlines can be computed through numerical integration of the horizontal velocity vector (Eq. 7.12). Pathlines can be traced against the flow in the same fashion by starting at the well and integrating the negative velocity vector (i.e., integrating against the flow).

The following example concerns two wells in an otherwise uniform flow field in the x-direction. The discharge of both wells is equal to Q. The 5-year capture zone is computed by starting 20 pathlines equally distributed along the circumference of each well. The pathlines are continued for 5 years against the flow (Figure 7.5).

```
# additional parameters
xw0, yw0 = 0, 0 # x,y-location well 0, m
xw1, yw1 = 100, 100 # x,y-location well 1, m
```

```
# solution and plot
def minvxvy(t, xy): # function vxvy from Section 7.1
    return -vxvy(t, xy, xw=xw0, yw=yw0) - vxvy(t, xy, xw=xw1, yw=yw1)

from scipy.integrate import solve_ivp
xstart = rw * np.cos(np.arange(0, 2 * np.pi, np.pi / 10)) # spaced around well
ystart = rw * np.sin(np.arange(0, 2 * np.pi, np.pi / 10)) # spaced around well
plt.subplot(111, aspect=1)
for i in range(len(xstart)):
    path = solve_ivp(minvxvy, (0, 5 * 365), [xstart[i] + xw0, ystart[i] + yw0],
    t_eval=np.linspace(0, 5 * 365, 100))
    plt.plot(path.y[0], path.y[1], 'C1')
    path = solve_ivp(minvxvy, (0, 5 * 365), [xstart[i] + xw1, ystart[i] + yw1],
    t_eval=np.linspace(0, 5 * 365, 100))
    plt.plot(path.y[0], path.y[1], 'C2')
```

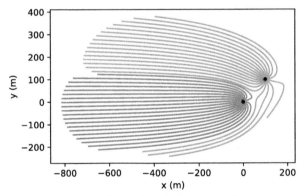

Figure 7.5 Capture zones for two wells with equal discharge in a uniform background flow; pathlines are obtained through numerical integration.

Exercise 7.5: Compare the analytic solution of the 5-year capture zone for a single well in an otherwise uniform flow field to the capture zone obtained with the numerical procedure.

7.3   A well in uniform background flow near a river

The discharge potential and stream function for groundwater flow problems governed by Laplace's equation (i.e., steady flow and negligible areal infiltration) may be combined into a complex potential, because the potential and stream function fulfill the Cauchy–Riemann conditions (Eq. 6.17). The real part of the complex potential Ω is the discharge potential, and the imaginary part is the stream function

$$\Omega = \Phi + i\Psi \tag{7.17}$$

In mathematical terms, this means that the stream function is the harmonic conjugate of the discharge potential (Strack, 1989). The complex potential Ω is a function of the complex coordinate $\zeta = x + iy$, where i is the imaginary unit ($i^2 = -1$). The letter j is used for the imaginary unit in Python, but i is used in the mathematical equations in this book. The negative derivative of the complex potential is the complex discharge

$$W(\zeta) = -\frac{d\Omega}{d\zeta} = Q_x - iQ_y \tag{7.18}$$

Note that the imaginary part of the complex discharge is minus the component of the discharge vector in the y-direction.

The introduction of complex variables makes it easy to compute the discharge potential and stream function simultaneously and allows for the use of sophisticated tools, including conformal mapping (see Chapter 10), to solve groundwater flow problems. More information on the application of complex variables in solving groundwater flow problems can be found in, e.g., Polubarinova-Kochina (1962), Harr (1962), Verruijt (1970), and Strack (1989).

The complex potential for uniform flow is

$$\Omega = -W_u\zeta \tag{7.19}$$

where W_u is the uniform complex discharge

$$W = W_u = U - iV \tag{7.20}$$

The complex potential for a well is

$$\Omega = \frac{Q}{2\pi} \ln(\zeta - \zeta_w) \tag{7.21}$$

where Q is the discharge of the well and $\zeta_w = x_w + iy_w$ is the complex coordinate of the well location. The validity of Eq. (7.21) is shown by writing $\zeta - \zeta_w$ in polar form

$$\zeta - \zeta_w = re^{i\theta} \tag{7.22}$$

where r is the distance between ζ and ζ_w and θ is the angle that a vector pointing from ζ_w to ζ makes with the positive x-direction. Substitution of Eq. (7.22) for $\zeta - \zeta_w$ in Eq. (7.21) gives

$$\Omega = \frac{Q}{2\pi} \ln(re^{i\theta}) = \frac{Q}{2\pi} \ln(r) + i\frac{Q}{2\pi}\theta \tag{7.23}$$

which means that the real part of Ω is indeed equal to the discharge potential for a well (Eq. 6.37) and the imaginary part is equal to the stream function (Eq. 6.41). The corresponding complex discharge for a well is

$$W = -\frac{d\Omega}{d\zeta} = -\frac{Q}{2\pi}\frac{1}{\zeta - \zeta_w} \tag{7.24}$$

The complex potential for a well in uniform background flow is obtained through superposition of Eqs. (7.19) and (7.21) as

$$\Omega = -W_u\zeta + \frac{Q}{2\pi} \ln(\zeta - \zeta_w) + C \tag{7.25}$$

where an arbitrary constant C is added for completeness. The real part of Ω is equivalent to the discharge potential (Eq. 7.5), and the imaginary part of Ω is equivalent to the stream function (Eq. 7.6).

As an example, a flow net for the problem of Figure 7.1 is created using the complex potential and shown in Figure 7.6. Evaluation of the stream function with Eq. (7.21) in Python results in the regular branch cut extending along the negative real axis so that the stream function is equal to $\Psi = 0$ along the capture zone envelope.

Although a technicality, the branch cut remains an ugly side effect of the use of the stream function. Alternatively, the logarithm in Eq. (7.21) may be defined as $\log(\zeta_w - \zeta)$ which results in the angle θ to be zero along the negative x-axis and a branch cut that jumps from π to $-\pi$ along the positive x-axis when going in the counterclockwise direction. This may be preferable in the visualization of some flow nets. Both options are shown in the example below (Figure 7.6). For the latter choice of the branch cut, the stream function equals $\Psi = -Q/2$ along the upper part of the capture zone envelope, and $\Psi = Q/2$ along the lower part of the capture zone envelope.

```
# parameters
Q = 80 # discharge of well, m^3/d
Wu = 0.5 - 0j # complex uniform flow, m^2/d
zetaw = 0 + 0j # complex location of well, m
```

```
# solution
xg, yg = np.meshgrid(np.linspace(-200, 100, 100), np.linspace(-100, 100, 100))
zetag = xg + yg * 1j # grid of complex coordinates
omega1 = -Wu * zetag + Q / (2 * np.pi) * np.log(zetag - zetaw)
omega2 = -Wu * zetag + Q / (2 * np.pi) * np.log(zetaw - zetag)
```

```
# basic plot of flow net
plt.subplot(121, aspect=1)
plt.contour(xg, yg, omega1.real, colors='C0')
plt.contour(xg, yg, omega1.imag, colors='C1', linestyles='-')
plt.contour(xg, yg, omega1.imag, [0], colors='k')
plt.subplot(122, aspect=1)
plt.contour(xg, yg, omega2.real, colors='C0')
plt.contour(xg, yg, omega2.imag, colors='C1', linestyles='-')
plt.contour(xg, yg, omega2.imag, [-40, 40], colors='k');
```

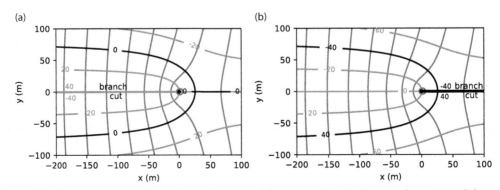

Figure 7.6 Flow net for a well in uniform background flow created with the complex potential; branch cut extending along negative x-axis (left) and along the positive x-axis (right); contour interval $\Delta\Psi = 20$ m^3/d

Exercise 7.6: Repeat Exercise 7.4, but with complex variables.

Exercise 7.7: Consider two extraction wells with equal discharge Q. One well is located at $(x, y) = (0, -50)$, and the other well, at $(x, y) = (0, 50)$. The two wells are located in an otherwise uniform flow from west to east; the gradient far away is $\partial h/\partial x = -0.002$. The transmissivity of the aquifer is $T = 200 \text{ m}^2/\text{d}$. Compute the minimum discharge Q_{min} so that no part of the background uniform flow passes between the two wells. Note that there are two stagnation points on the x-axis when $Q > Q_{min}$. Draw the capture zone envelopes for two cases: $Q = Q_{min}$ and $Q = 200 \text{ m}^3/\text{d}$.

Pumping wells that are located near rivers intercept groundwater that is on its way to the river. When the discharge of the pumping well is large enough, part of the water pumped from the well may originate from the river or from the other side of the river. Either way, the flow in the river is depleted. In this section, wells are considered that are pumping near long and straight river sections, while the background flow is uniform and toward the river on both sides of the river. The aquifer is confined and has transmissivity T. A well with discharge Q is located at $(x, y) = (-d, 0)$, and the river runs along the y-axis. The river stage is h_0, which corresponds to a potential $\Phi_0 = Th_0$. The river is in direct contact with the aquifer so the boundary condition along the y-axis is

$$\Phi|_{x=0,y} = \Phi_0 \tag{7.26}$$

The background flow is uniform and normal to the river, but potentially different on either side of the river

$$Q_x = U_L \qquad x \le 0 \tag{7.27}$$
$$Q_x = U_R \qquad x \ge 0 \tag{7.28}$$

Note that because the river is in direct contact with the aquifer, the flow on the right side of the river remains uniform (Eq. 7.28) even if there is a well on the left side. In the next section, the bottom of the river is formed by a leaky stream bed.

The solution for the complex potential is obtained through the method of images and superposition as

$$\Omega = -U_L\zeta + \frac{Q}{2\pi} \ln\left(\frac{\zeta+d}{d-\zeta}\right) + \Phi_0 \qquad x \le 0 \tag{7.29}$$

and the corresponding complex discharge is

$$W = -\frac{d\Omega}{d\zeta} = U_L - \frac{Q}{2\pi}\left(\frac{1}{\zeta+d} - \frac{1}{\zeta-d}\right) \qquad x \le 0 \tag{7.30}$$

The critical discharge of the well is defined as the maximum discharge at which the well doesn't draw any water from the river into the aquifer. When the well pumps at the critical discharge, the discharge vector at the point on the river closest to the well equals zero so that the point $(x, y) = (0, 0)$ is a stagnation point. An equation for the stagnation point(s) in the flow field may be obtained by setting $W = 0$ (Eq. 7.30) and solving for ζ, which gives the complex locations ζ_s of the stagnation points as

$$\zeta_s = \pm\sqrt{d^2 - \frac{Qd}{\pi U_L}} \tag{7.31}$$

According to this equation, there are always two stagnation points, except for the case that the argument of the square root equals zero, which means the stagnation point is at the origin. The corresponding critical discharge Q_{cr} is

$$Q_{cr} = \pi U_L d \tag{7.32}$$

When $Q < Q_{cr}$, the well does not capture any river water and there are two stagnation points located on the x-axis, one on the side of the river where the well is located and one on the other (image) side. When $Q > Q_{cr}$, there are two imaginary stagnation points located on the river (the y-axis). Water infiltrates into the aquifer between the stagnation points on the river, and this water flows toward the well.

In the example below, the discharge of the well is $Q_{cr}/2$ for the left flow net and $2Q_{cr}$ for the right flow net (Figure 7.7). The flow net is shown on both sides of the river; flow is uniform on the right side of the river and not altered by the well (because the river is in direct contact with the aquifer and the Dupuit–Forchheimer approximation is adopted). Note that there is only one stagnation point in the flow field for the case that $Q = Q_{cr}/2$; the other stagnation point is at the image location. In the code below, the square root in Eq. (7.31) is computed with the sqrt function of the numpy.lib.scimath package, which automatically returns a complex number when the argument of the square root is negative. The amount ΔQ of the well discharge that originates from the river is obtained by subtracting the values of the stream function at the stagnation points on the river.

```
# parameters
dhdx = -0.001 # head gradient, -
h0 = 10 # water level in river, m
T = 100 # transmissivity, m^2/d
UL = -T * dhdx # uniform flow on left side of river, m^2/d
UR = -UL # uniform flow on right side of river, m^2/d
d = 50 # distance between well and river, m
phi0 = T * h0 # potential along the river, m^3/d
Qcr = np.pi * UL * d # critical discharge, m^3/d
print(f'{Qcr = :.2f} m^3/d')
```

```
Qcr = 15.71 m^3/d
```

```
# solution
def omega(x, y, Q, d=d, phi0=phi0):
    zeta = x + y * 1j
    if x <= 0:
        om = -UL * zeta + Q / (2 * np.pi) * \
            np.log((zeta + d) / (d - zeta)) + phi0
    else:
        om = -UR * zeta + phi0
    return om

omegavec = np.vectorize(omega) # vectorized version of omega function

from numpy.lib.scimath import sqrt as csqrt
xg, yg = np.meshgrid(np.linspace(-200, 50, 200), np.linspace(-100, 100, 200))
om1 = omegavec(xg, yg, Q=Qcr / 2) # Q = half critical discharge
om2 = omegavec(xg, yg, Q=2 * Qcr) # Q = twice critical discharge
zetas = csqrt(d ** 2 - 2 * Qcr * d / (np.pi * UL)) # Q = 2Qcr
ys = zetas.imag # positive y-location of stagnation point
delQ = omega(0, ys, 2 * Qcr).imag - omega(0, -ys, 2 * Qcr).imag
print(f'fraction of water from river for Q = 2Q_cr: {delQ / (2 * Qcr):.2f}')
```

```
fraction of water from river for Q = 2Q_cr: 0.18
```

```
# basic flow nets
plt.subplot(121, aspect=1)
```

```
plt.contour(xg, yg, om1.real, np.arange(phi0 - 30, phi0 + 30, Qcr / 8))
plt.contour(xg, yg, om1.imag, np.arange(-2 * Qcr, 3 * Qcr, Qcr / 8))
plt.subplot(122, aspect=1)
plt.contour(xg, yg, om2.real, np.arange(phi0 - 30, phi0 + 30, Qcr / 8))
plt.contour(xg, yg, om2.imag, np.arange(-2 * Qcr, 3 * Qcr, Qcr / 8));
```

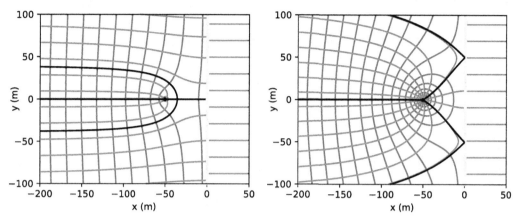

Figure 7.7   Flow net for a well in uniform background flow near a river with zero bed resistance, $Q = Q_{cr}/2$ (left) and $Q = 2Q_{cr}$ (right); capture zone envelopes are shown in black.

7.4   A well in uniform background flow near a river with a leaky stream bed

The solution for a well in a uniform background flow near a long and straight river segment in direct contact with the aquifer is modified to include a leaky stream bed below the river. All boundary conditions are the same as for the solution in the previous section, except for the boundary condition along the river. The river has a leaky stream bed so that there is a hydraulic resistance between the river and the aquifer (Figure 7.8). Therefore, the pumping well on one side of the river results in a drawdown on the other side of the river, unlike in the previous section. The pumping well may capture water from the other side of the river, which can have major implications for a safe groundwater supply, e.g., when there is a contamination on the other side of the river (e.g., Harr, 1996).

The river has a small width B, and the stream bed has a resistance to vertical flow c. Leakage to or from the river is approximated as occurring along the centerline of the stream, and the width of the stream is neglected in the solution. This approximation may be reasonable when the stream is narrow relative to the aquifer thickness and the leakage factor ($\sqrt{kHc}$). The vertical specific discharge through the leaky stream bed may be written as

$$q_z = \frac{h - h_0}{c} \tag{7.33}$$

The inflow σ [L²/T] from the aquifer into the river per unit length of river is obtained as

$$\sigma(y) = Bq_z = B\frac{h|_{x=0,y} - h_0}{c} = \frac{C}{T}(\Phi|_{x=0,y} - \Phi_0) \tag{7.34}$$

where the conductance of the stream bed is defined as $C = B/c$. The boundary condition along the river is obtained from continuity of flow as

$$\sigma = Q_x|_{x=0^-,y} - Q_x|_{x=0^+,y} = \frac{C}{T}(\Phi|_{x=0,y} - \Phi_0) \tag{7.35}$$

where $x = 0^-$ and $x = 0^+$ are the x-locations just to the left and right of the centerline of the river, respectively. The second boundary condition along the river is that the head in the aquifer on the left side of the river is equal to the head in the aquifer on the right side of the river, or in terms of the discharge potential (recall that the width B of the river is neglected in the solution)

$$\Phi|_{x=0^-,y} = \Phi|_{x=0^+,y} \tag{7.36}$$

The uniform background flow on both sides of the river is the same as Eqs. (7.27) and (7.28), as shown in Figure 7.8.

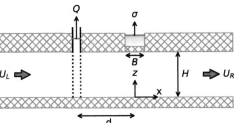

Figure 7.8 Vertical cross section of a well near a river with a leaky stream bed.

The complex potential solution for the stated problem is obtained from Bakker and Anderson (2003). The complex potential is

$$\Omega = \frac{Q}{2\pi} \ln \frac{\zeta + d}{d - \zeta} - \frac{Q}{2\pi} \exp(Z_0) E_1(Z_0) - U_L\zeta + A \qquad x \le 0 \tag{7.37}$$

$$\Omega = -\frac{Q}{2\pi} \exp(Z_1) E_1(Z_1) - U_R\zeta + A \qquad x \ge 0 \tag{7.38}$$

where

$$Z_0 = -\frac{C}{2T}(\zeta - d) \tag{7.39}$$

$$Z_1 = \frac{C}{2T}(\zeta + d) \tag{7.40}$$

$$A = \frac{T}{C}(U_L - U_R) + \Phi_0 \tag{7.41}$$

and E_1 is the exponential integral, which is available as exp1 in the scipy.special package. The corresponding complex discharge vector is

$$W = -\frac{Q}{2\pi}\left[\frac{1}{\zeta + d} + \frac{C}{2T}\exp(Z_0)E_1(Z_0)\right] + U_L \qquad x \le 0 \tag{7.42}$$

$$W = -\frac{Q}{2\pi}\left[\frac{1}{\zeta + d} - \frac{C}{2T}\exp(Z_1)E_1(Z_1)\right] + U_R \qquad x \ge 0 \tag{7.43}$$

The complex potential is implemented in a Python function in the code below.

```
# parameters
T = 100 # transmissivity, m^2/d
h0 = 10 # water level in river, m
UL = 0.1 # uniform flow on left side of river, m^2/d
UR = -0.1 # uniform flow on right side of river, m^2/d
d = 50 # distance between well and river, m
```

```
phi0 = T * h0 # potential along the river, m^3/d
rw = 0.3 # radius of well, m
kstreambed = 0.02 #hydraulic conductivity of stream bed, m/d
Hstreambed = 0.2 #thickness of stream bed, m
B = 5 # width of streambed, m
c = Hstreambed / kstreambed #  resistance of stream bed, d
C = B / c #conductance of stream bed, m/d
print(f'leakage factor below stream: {np.sqrt(T * c):.0f} m')
```

```
leakage factor below stream: 32 m
```

```
# solution
from scipy.special import exp1

def omegaresriv(x, y, Q):
    A = T / C * (UL - UR) + phi0
    z = x + y * 1j
    if x <= 0:
        Z0 = -C / (2 * T) * (z - d)
        om = Q / (2 * np.pi) * (np.log((z + d) / (d - z)) -
            np.exp(Z0) * exp1(Z0)) - UL * z + A
    else:
        Z1 = C / (2 * T) * (z + d)
        om = -Q / (2 * np.pi) * np.exp(Z1) * exp1(Z1) - UR * z + A
    return om

omegaresrivvec = np.vectorize(omegaresriv)
```

The critical discharge of the well is determined such that the well captures water only from the uniform background flow on the side of the river where the well is located. The critical discharge occurs, as for the case with no stream bed resistance (Section 7.3), when the origin is a stagnation point. The critical discharge cannot be determined analytically in this case, but is found numerically. A function func is written that returns the x-component of the discharge vector at $(x, y) = (0^-, 0)$ for a given value of Q. The brentq method of the scipy.optimize package is used to find the value of the discharge such that the function func equals zero, which corresponds to the critical discharge; the brentq method is an efficient and robust method to find the zero (the root) of a function in a bracketing interval. In the example below, a relatively small resistance of the stream bed is used, which already increases the critical discharge of the well by a significant amount as compared to the solution with zero stream bed resistance (Section 7.3).

```
# find critical discharge
def Qxriverleft(Q, y):
    z = -1e-6 + y * 1j
    Z0 = -C / (2 * T) * (z - d)
    W = -Q / (2 * np.pi) * (1 / (z + d) +
                    C / (2 * T) * np.exp(Z0) * exp1(Z0)) + UL
    return W.real

def func(Q):
    return Qxriverleft(Q, 0)

from scipy.optimize import brentq
Qcr2 = brentq(func, 0, 50) # critical discharge with bed resistance
```

```
Qcr1 = np.pi * UL * d # critical discharge with no bed resistance

print(f'River bed conductance C = {C:.1f} m^2/d')
print(f'Qcr without bed resistance = {Qcr1:.2f} m^3/d')
print(f'Qcr with bed resistance = {Qcr2:.2f} m^3/d')

River bed conductance C = 0.5 m^2/d
Qcr without bed resistance = 15.71 m^3/d
Qcr with bed resistance = 25.54 m^3/d
```

A contour plot is produced for the case that the well pumps at the critical discharge without stream bed resistance (the solution from the previous section) and for the case with stream bed resistance (Figure 7.9). The contour interval is different for both cases. Note that for the case with bed resistance, the flow on the right side of the river is affected by the well pumping on the left side of the river. This is more pronounced when the discharge of the well is increased in the following example.

```
xg, yg = np.meshgrid(np.linspace(-200, 50, 100), np.linspace(-100, 100, 100))
om1 = omegavec(xg, yg, Q=Qcr1) # omegavec from previous section
om2 = omegaresrivvec(xg, yg, Q=Qcr2)
```

```
# basic flow nets
plt.subplot(121, aspect=1)
plt.contour(xg, yg, om1.real, np.arange(phi0 - 30, phi0 + 30, Qcr1 / 8))
plt.contour(xg, yg, om1.imag, np.arange(-2 * Qcr1, 3 * Qcr1, Qcr1 / 8))
plt.axvline(0, color='lightblue', lw=3)
plt.subplot(122, aspect=1)
plt.contour(xg,
yg, om2.real,
np.arange(phi0 - 60, phi0 + 60, Qcr2 / 8))
plt.contour(xg, yg, om2.imag, np.arange(-2 * Qcr2, 3 * Qcr2, Qcr2 / 8))
plt.axvline(0, color='lightblue', lw=3);
```

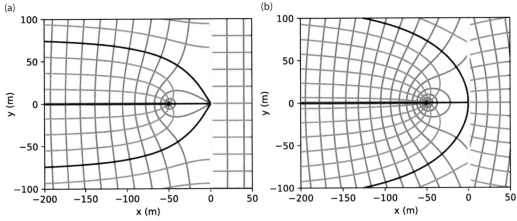

Figure 7.9 Flow net for a well pumping at critical pumping rate in uniform background flow near a river with zero bed resistance ($C = \infty$, left) and non-zero bed resistance ($C = 0.5$ m^2/d, right); the capture zone envelopes are shown in black.

When the discharge of the well is larger than Q_{cr}, the well may draw water from the river, or it may draw water from the other side of the river, or both. This means that there is a section of river near the well where water on the left side of the river flows toward the well.

The points on the river where the flow in the aquifer changes from toward the river to away from the river are obtained numerically. The function `func2` returns the Q_x component of the discharge vector just on the left side of the river for given y and Q values. The `brentq` function is used again to find the change point where `func2` equals zero. Note that for this case with river bed resistance, the change point is not a stagnation point, as Q_y does not equal zero. The change point is computed for the case that $Q = 2Q_{cr}$ as follows.

```
# changed parameters
Q = 2 * Qcr2

# solution
def func2(ys, Q):
    return Qxriverleft(Q, ys)

ys = brentq(func2, 0, 2 * d, args=(Q))
print(f'point on the river where Qx equals zero: {ys:.2f} m')
```

```
point on the river where Qx equals zero: 57.44 m
```

The corresponding flow net is shown in Figure 7.10. The code to produce the flow net is similar to the code for Figure 7.9 and is not repeated. The capture zone extends to the right side of the river. Whether the well captures water from the river or from the right side of the river is determined by the head in the aquifer at the river. If the head in the aquifer at the river is above the water level in the river ($h_0 = 10$ m), then water flows from the aquifer into the river. If the head in the aquifer at the river is below the water level in the river, then water flows from the river into the aquifer. For the values used in this example, the head in the aquifer at the origin is above the water level in the river so that the well captures water from the other side of the river, but not from the river itself.

```
print(f'head at origin: {omegaresriv(0, 0, Q).real / T:.2f} m')
print(f'head at well: {omegaresriv(-d + rw, 0, Q).real / T:.2f} m')
```

```
head at origin: 10.25 m
head at well: 9.87 m
```

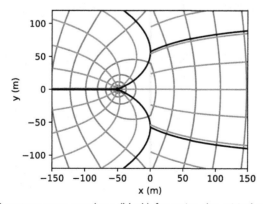

Figure 7.10 Flow net and capture zone envelope (black) for twice the critical pumping rate and $C = 0.5$ m^2/d. Note that the well draws water from the right side of the river.

Exercise 7.8: Consider the example with stream bed resistance presented above. Compute the head along the river for the case that the discharge of the well equals zero.

7.5   A well in uniform background flow near the coast

The solutions in the previous sections concerned the capture zone of a well in uniform background flow near a straight river. The critical discharge of the well was derived such that the well does not capture water from the river or from the other side of the river. In this section, the straight river is replaced by a straight and vertical coastal boundary and the maximum discharge of the well is determined such that the well does not pump any saltwater. In other words, the maximum discharge is determined such that the interface does not reach the well. It turns out that this maximum discharge is quite a bit smaller than the maximum discharge for a well near a river (Section 7.3).

Consider a well in a coastal aquifer. The coastline is located along the line $x = 0$. The well is located at $(x, y) = (-d, 0)$, has a discharge Q, and has a radius r_w. In the absence of the well, the flow toward the coast is uniform and equal to $Q_x = U$. The aquifer is unconfined, the hydraulic conductivity is k, and the elevation of the bottom of the aquifer is z_b with respect to sea level. The density of the seawater is ρ_s. The sea is in direct contact with the aquifer (no seabed resistance) so that the head along the coastline is equal to sea level. The solution for the complex potential is identical to the solution for a well near a river in direct contact with the aquifer (Eq. 7.29). For this case of unconfined interface flow, the potential at the toe of the interface is Eq. (4.26)

$$\Phi_{\text{toe}} = \tfrac{1}{2} k \frac{\alpha + 1}{\alpha^2} z_b^2 \tag{7.44}$$

In the absence of pumping, the toe of the interface is located at (Eq. 4.21)

$$x_{\text{toe}} = -\frac{\Phi_{\text{toe}}}{U} \tag{7.45}$$

First, consider the case that the discharge of the well is half the critical discharge for a well near a river (Eq. 7.32). A flow net is constructed showing that indeed the stagnation point is still far from the coast line for this value of Q (Figure 7.11). The toe of the interface in the absence of pumping is plotted with a black dashed line, while the contour line representing Φ_{toe} for the case with pumping is plotted with a black solid line. It turns out that there are two contour lines where the potential is equal to Φ_{toe}. One contour line runs from north to south and curves toward the well: This is the location of the toe of the interface during pumping. A second black contour is present around the well. This is because the discharge potential at the well is below Φ_{toe}, but this does not mean that the well pumps saltwater. The black contour line around the well is not connected to the black contour line near the coast, which means that there is no way for the saltwater to reach the well. This is further illustrated in the right graph of Figure 7.11, where the potential is plotted as a function of x along the line $y = 0$. When converting the potential to a head, the values of the potential inside the closed black contour around the well must be converted to a head using the formula for unconfined flow rather than for unconfined interface flow.

```
# parameters
rw = 0.3 # radius of well, m
k = 40 # hydraulic conductivity, m/d
zb = -20 # elevation of bottom of aquifer, m
rhof = 1000 # density of freshwater, kg/m^3
rhos = 1025 # density of saltwater, kg/m^3
U = 1 # uniform flow in x-direction, m^2/d
d = 1000 # distance of well from coast line, m
Qcr = np.pi * U * d # critical discharge for well near river, m^3/d
```

```
Q = Qcr / 2 # discharge of well, m^3/d
print(f'discharge of well: {Q:.2f} m^3/d')
```

discharge of well: 1570.80 m^3/d

```
# solution
alpha = rhof / (rhos - rhof) # alpha factor
phitoe = 0.5 * k * (alpha + 1) / alpha ** 2 * zb ** 2
xtoe = -phitoe / U # toe in the absence of pumping

def omega(x, y, Q):
    zeta = x + y * 1j
    om = Q / (2 * np.pi) * np.log((zeta + d) / (d - zeta)) - U * zeta
    return om

xg, yg = np.meshgrid(np.linspace(-1200, 0, 120), np.linspace(-500, 500, 101))
om = omega(xg, yg, Q)
```

```
# basic flow net and potential along y=0
plt.subplot(121, aspect=1)
plt.contour(xg, yg, om.real, np.arange(-2000, 2000, 100), colors='C0')
plt.contour(xg, yg, om.real, [phitoe], colors='k')
plt.contour(xg, yg, om.imag, np.arange(-2000, 2000, 100), colors='C1')
plt.axvline(xtoe, color='k', linestyle='--')
plt.subplot(122) # potential along y=0
plt.plot(xg[0], om.real[50], label=r'$\Phi(x,0)$')
plt.axhline(phitoe, ls='--', label=r'$\Phi_{toe}$')
plt.legend();
```

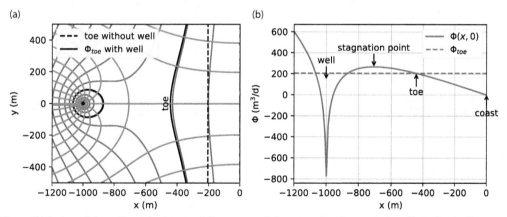

Figure 7.11 A well in uniform background flow toward the coast for the case that $Q < Q_{max}$; flow net (left) and potential along $y = 0$ (right).

The maximum discharge Q_{max} of the well is reached when the two black contours for Φ_{toe} touch. This happens when the potential at the stagnation point is equal to Φ_{toe}. The well starts pumping saltwater when the discharge is larger than Q_{max}, and the presented solution is not valid anymore, as the solution is only valid for stagnant saltwater. This is in contrast to the solutions presented in the previous two sections where the solutions were equally valid for well discharges below and above the critical pumping rate Q_{cr}.

The maximum discharge is found numerically. A function is written that takes the discharge Q as input and returns the difference between the discharge potentials at the stagnation point and Φ_{toe}. This function equals zero for the maximum discharge. The `brentq` method is used to find Q_{max}. Note that the maximum discharge is just slightly more than half the critical discharge for a well near a river (Eq. 7.32).

```
def func(Q):
    xs = -np.sqrt(d ** 2 - Q * d / (np.pi * U))
    return omega(xs, 0, Q).real - phitoe

from scipy.optimize import brentq
Qmax = brentq(func, 1, np.pi * U * d)
print(f'maximum discharge: {Qmax:.2f} m^3/d')
print(f'Qmax as fraction of Qcr of Section 7.3: {Qmax / Qcr:.2f}')
om1 = omega(xg, yg, Qmax)
```

```
maximum discharge: 1803.22 m^3/d
Qmax as fraction of Qcr of Section 7.3: 0.57
```

When pumping at slightly below the maximum discharge, the toe of the interface is still at a significant distance from the well. When the discharge is increased to the maximum discharge, the toe of the saltwater suddenly breaks through to the well as the potential at the stagnation point drops below the potential at the toe; that is, the head at the stagnation point drops below $-z_b/\alpha$. A flow net for the maximum discharge is shown in Figure 7.12; the code is the same as that for Figure 7.11 and is not repeated. The potential along the line $y = 0$ shows that the maximum potential between the well and the coast is indeed Φ_{toe}.

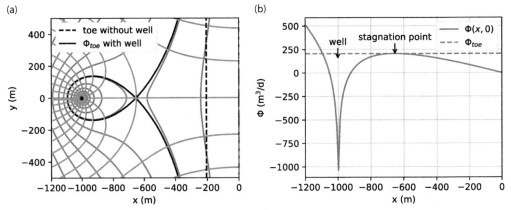

Figure 7.12 A well in uniform background flow toward the coast for the case that $Q = Q_{max}$; flow net (left) and potential along $y = 0$ (right).

Exercise 7.9: The head and interface position along the line $y = 0$ for $Q = Q_{max}$ are shown in Figure 7.13. Compute the head at the well for this case.

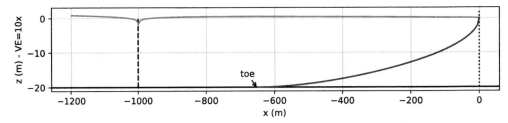

Figure 7.13 The head and interface position for $Q = Q_{max}$ along the line $y = 0$.

Chapter 8

Analytic element modeling of steady two-dimensional flow

Pumping wells form an important part of the two-dimensional solutions that have been discussed so far. The principle of superposition allows for the solution of problems with an arbitrary number of wells, and the method of images allows for the simulation of straight specified-head and no-flow boundaries. The organization of solutions with multiple wells becomes cumbersome when the number of wells grows. The analytic element method is a modeling technique that facilitates the superposition of many solutions, including wells.

The analytic element method is based on the superposition of analytic solutions (e.g., Strack, 2003). Each analytic solution represents a feature in the aquifer and has at least one free parameter. For example, the solution for a pumping well is an analytic element and the free parameter is the discharge of the well. The free parameter of an analytic element may either be specified or computed from specified conditions. For example, the discharge of a well may be computed from the head specified at a point, referred to as the collocation point. All unknown free parameters in a model must be computed simultaneously as, e.g., the discharge of one well may affect the head at another well and vice versa.

In this chapter, the principles of the analytic element are discussed. Python code for the analytic element method is developed using an object-oriented design based on Bakker and Kelson (2009). The analytic element program presented in this chapter includes uniform flow, wells, line-sinks for the modeling of river segments, and area-sinks for the modeling of areal recharge. Many advanced analytic elements exist for the simulation of steady single-layer flow, e.g., for the simulation of impermeable boundaries or inhomogeneity boundaries (e.g., Haitjema, 1995; Strack, 2017). In addition, formulations exist for both steady and transient multi-layer flow (Bakker and Strack, 2003; Fitts, 2010; Bakker, 2013a). Computer programs for analytic element modeling of groundwater flow include Gflow (Haitjema, 1995) for steady single layer flow, and TimML (Bakker, 2021), TTim (Bakker, 2013b), and AnAqSim (Fitts, 2021) for steady and transient multi-aquifer flow.

8.1 Uniform flow and wells

In this section, an analytic element code is developed to simulate wells, uniform flow, and a constant. The complex potential is written as

$$\Omega = -Q_u e^{-i\theta} \zeta + \sum_{m=1}^{M} \frac{Q_m}{2\pi} \ln(\zeta - \zeta_m) + \sum_{n=1}^{N} \frac{P_n}{2\pi} \ln(\zeta - \zeta_n) + C \tag{8.1}$$

where Q_u is the magnitude of the uniform flow (the parameter), θ is the direction of uniform flow, M is the number of wells with a specified discharge, N is the number of wells with a head-specified condition, and P_n is the discharge of head-specified well n, which is unknown

DOI: 10.1201/9781315206134-8

upfront. The real constant C is also a priori unknown and is computed from a specified-head condition. The transmissivity is approximated as constant throughout the model.

The analytic element code is developed using object-oriented programming; an introduction to object-oriented programming for Python may be found in, e.g., Lutz (2013). The analytic element code consists of a Model class and a number of Element classes. The Model class stores the transmissivity of the aquifer and a Python list called elements which contains all the elements in the model; the element list is initially empty. The Model class has a function omega to compute the complex potential at a point caused by all analytic elements in the model (Eq. 8.1). In addition, the Model class has separate potential, streamfunction, and head functions. Finally, the Model class has a solve method, which computes the parameters of all head-specified elements in the model. This function is included here, but will be explained later on.

```python
class Model:

    def __init__(self, T):
        self.T = T
        self.elements = []  # list of all elements

    def omega(self, x, y):
        omega = 0 + 0j
        for e in self.elements:
            omega += e.omega(x, y)
        return omega

    def potential(self, x, y):
        return self.omega(x, y).real

    def streamfunction(self, x, y):
        return self.omega(x, y).imag

    def head(self, x, y):
        return self.potential(x, y) / self.T

    def solve(self):
        esolve = [e for e in self.elements if e.nunknowns == 1]
        nunknowns = len(esolve)
        matrix = np.zeros((nunknowns, nunknowns))
        rhs = np.zeros(nunknowns)
        for irow in range(nunknowns):
            matrix[irow], rhs[irow] = esolve[irow].equation()
        solution = np.linalg.solve(matrix, rhs)
        for irow in range(nunknowns):
            esolve[irow].parameter = solution[irow]
```

The Element class is the base class of all analytic elements. It stores the model that the element is added to, the value of the parameter (e.g., the discharge of a well), and the number of unknown parameters of an element, which is zero by default, and it adds the element to the specified model by adding it to the elements list of the model. The Element class has three functions that are the same for all elements: omega, potential, and potinf. The omega function computes the complex potential of an element by returning the product of the influence function omegainf of the element and the parameter of the element. Every element must implement its own omegainf, the complex potential influence function, which is the complex potential for the case that the parameter of the element equals one. The potential function returns the discharge potential of an element, the real part of `omega`. The `potinf` function returns the influence function of the discharge potential, the real part of the `omegainf` function. The utility of influence functions will become apparent further on in this section, when elements are introduced that have an a priori unknown parameter, for example a well where the discharge is unknown and is computed from a specified head at the well.

```
class Element:

    def __init__(self, model, p):
        self.model = model
        self.parameter = p
        self.nunknowns = 0
        self.model.elements.append(self)

    def omega(self, x, y):
        return self.parameter * self.omegainf(x, y)

    def potential(self, x, y):
        return self.omega(x, y).real

    def potinf(self, x, y):
        return self.omegainf(x, y).real
```

Every element is derived from the `Element` base class so that it automatically gets all attributes and functions from the `Element` class. The `Well` class, for wells with a specified discharge, takes as input arguments the model to which the well is added, the location of the well, the discharge of the well, and the radius of the well. The `Well` class is derived from the `Element` class, which stores the parameter (the discharge). The `Well` class stores the complex location and the radius of the well. Only one function needs to be implemented for the `Well` class: the `omegainf` function, which for a well is

$$\Omega = \frac{1}{2\pi} \ln(\zeta - \zeta_w) \qquad (8.2)$$

The value of $\zeta - \zeta_w$ is set to the radius of the well for points inside the radius of the well using the `np.where` function.

```
class Well(Element):

    def __init__(self, model, xw=0, yw=0, Q=1, rw=0.3):
        Element.__init__(self, model, p=Q)
        self.zetaw = xw + 1j * yw
        self.rw = rw

    def omegainf(self, x, y):
        zminzw = x + 1j * y - self.zetaw
        zminzw = np.where(np.abs(zminzw) < self.rw, self.rw, zminzw)
        return 1 / (2 * np.pi) * np.log(zminzw)
```

The `UniformFlow` class is also derived from the `Element` class and takes as input arguments the model to which it is added, the absolute value of the head gradient, and the direction θ of uniform flow (in degrees measured counterclockwise from the positive x-direction). The `omegainf` function is

$$\Omega = -e^{-i\theta}\zeta \qquad (8.3)$$

```
class UniformFlow(Element):

    def __init__(self, model, gradient, angle):
        Element.__init__(self, model, p=model.T * gradient)
        self.udir = np.exp(-1j * np.deg2rad(angle))

    def omegainf(self, x, y):
        return -self.udir * (x + y * 1j)
```

An analytic element model may now be constructed by creating a `Model` object, and option-ally adding a uniform flow object and an arbitrary number of wells. The head, potential, and stream function may be computed at any point. In the example below, one extraction well and one injection well are situated in an otherwise uniform flow field in the southeastern direction ($\theta = -45°$). A flow net is shown in Figure 8.1. All water from the injection well (which has a smaller discharge than the extraction well) is captured by the extraction well.

```
# parameters and model
ml = Model(T=100) # transmissivity, m^2/d
Well(ml, xw=50, yw=0, Q=200) # extraction well, Q in m^3/d
Well(ml, xw=0, yw=0, Q=-100) # injection well, Q in m^3/d
UniformFlow(ml, gradient=0.002, angle=-45);
```

```
# solution
xg, yg = np.meshgrid(np.linspace(-100, 100, 100), np.linspace(-75, 75, 100))
pot = ml.potential(xg, yg)
psi = ml.streamfunction(xg, yg)
```

```
# basic flow net
plt.subplot(111, aspect=1)
plt.contour(xg, yg, pot, 20)
plt.contour(xg, yg, psi, 20);
```

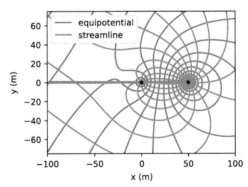

Figure 8.1 Flow net for analytic element model of an injection well ($Q = -100$ m³/d) and an extraction well ($Q = 200$ m³/d) in uniform flow in southeastern direction.

Exercise 8.1: Create a flow net for the example of a well in uniform flow as shown in Figure 7.1 (but with equipotential and streamlines) using the analytic element code developed above. Use a transmissivity of $T = 250$ m²/d.

The head must be defined at at least one point in the aquifer; otherwise, the head values along the contours are meaningless (an arbitrary value can be added to all values, resulting in the same flow field). Two types of head-specified elements are implemented here: wells with a specified head and an additive constant which is also computed from a specified-head condition. The free parameters of the head-specified elements are computed by building a system of equations for all head-specified conditions and solving for all free parameters simultaneously. This procedure is illustrated by solving a problem with uniform flow, two wells, and a constant.

Consider an aquifer with uniform flow U in the positive x-direction with two wells and a real constant C (Figure 8.2). The first well is located at $(x, y) = (0, 0)$ and has a specified discharge Q. The second well is located at $(x, y) = (a, b)$ and has a discharge P. The radius of both wells is r_w. The complex potential for this problem is obtained from superposition of the complex potential of uniform flow (Eq. 7.19), the complex potentials for two wells (Eq. 7.21), and a real constant C

$$\Omega = -U\zeta + \frac{Q}{2\pi}\ln(\zeta) + \frac{P}{2\pi}\ln(\zeta - (a + ib)) + C \tag{8.4}$$

The discharge Q of the first well is specified, but the discharge P of the second well is unknown and must be computed such that the head equals h_w at the second well. Furthermore, the head at $(x, y) = (a, 0)$ is known to be h_0. This gives two equations for the two unknown parameters C and P.

The parameters P and C are obtained from the system of two linear equations

$$\begin{aligned}
\Phi|_{\zeta=a} &= \Phi_0 \\
\Phi|_{\zeta=a+r_w+ib} &= \Phi_w
\end{aligned} \tag{8.5}$$

where Φ_0 and Φ_w are the potential equivalents of h_0 and h_w, respectively. Note that the head at the well is specified at $(x, y) = (a + r_w, b)$, which is referred to as the collocation point; specification at another point on the well screen may result in a slightly different solution, but the difference is commonly negligible. Substitution of the real part of the complex potential (Eq. 8.4) for Φ in Eq. (8.5) and rearrangement of terms gives the following system of two equations in two unknowns

$$\begin{aligned}
\frac{P}{2\pi}\ln(b) + C &= \Phi_0 + Ua - \frac{Q}{2\pi}\ln(a) \\
\frac{P}{2\pi}\ln(r_w) + C &= \Phi_w + U(a + r_w) - \frac{Q}{2\pi}\ln(d)
\end{aligned} \tag{8.6}$$

where $d = \sqrt{(a + r_w)^2 + b^2}$ is the distance between the well with specified discharge and the collocation point on the well screen of the head-specified well. In matrix form, the system of equations is

$$\begin{pmatrix} \frac{1}{2\pi}\ln(b) & 1 \\ \frac{1}{2\pi}\ln(r_w) & 1 \end{pmatrix} \begin{pmatrix} P \\ C \end{pmatrix} = \begin{pmatrix} \Phi_0 + Ua - \frac{Q}{2\pi}\ln(a) \\ \Phi_w + U(a + r_w) - \frac{Q}{2\pi}\ln(d) \end{pmatrix} \tag{8.7}$$

The unknown free parameters P and C are obtained by solving this system of two linear equations.

The solve method of the Model class, as defined above, counts the number of elements with an unknown parameter and stores them in a separate list called esolve, builds the matrix and the right-hand side of the system of equations, solves the system, and stores the computed parameters for the corresponding head-specified elements (in this case the values of P and C). The solve method loops through all elements in the esolve list and calls their equation function, which returns one row for the matrix and one right-hand side value. The equation method of the Constant element (not yet defined) returns the first row of the matrix and the first value of the right-hand side vector, and the equation method of the second well returns the second row of the matrix and the second value of the right-hand side vector.

The equation function is stored in a separate mix-in class HeadEquation, as it is the same for all head-specified elements. Each element that uses this mix-in class must store the location of the collocation point where the head is specified as xc and yc, and the specified value of the discharge potential at the collocation point as pc. The equation function starts out with a row of the matrix that is an empty list and a right-hand side value that is equal to the specified discharge potential pc. The equation function then loops through all the elements in the model. If an element has an unknown parameter (nunkonwns==1), its potential influence function potinf is evaluated at the collocation point and is stored as the next value in the row of the matrix. If an element does not have an unknown parameter, its potential function is evaluated at the collocation point and the returned value is subtracted from the right-hand side value.

```
class HeadEquation:

    def equation(self):
        row = []
        rhs = self.pc
        for e in self.model.elements:
            if e.nunknowns == 1:
                row.append(e.potinf(self.xc, self.yc))
            else:
                rhs -= e.potential(self.xc, self.yc)
        return row, rhs
```

The HeadWell class for head-specified wells can now be derived from both the Well class and the HeadEquation class. Input arguments are the model to which the element is added, the location of the well, the radius, and the specified head. The xc and yc locations of the collocation point are stored, as well as the specified potential pc at the collocation point. The number of unknowns parameters nunknowns is set to 1. All other attributes and functions are inherited from the Well and HeadEquation classes.

```
class HeadWell(Well, HeadEquation):

    def __init__(self, model, xw, yw, rw, hw):
        Well.__init__(self, model, xw, yw, 0, rw)
        self.xc, self.yc = xw + rw, yw
        self.pc = self.model.T * hw
        self.nunknowns = 1
```

The Constant class is derived from the Element class and the HeadEquation class. Input arguments are the model to which the element is added and the location and value of the specified head. The omegainf function of the Constant class is simply equal to 1. The Constant element is also referred to as the reference point.

```
class Constant(Element, HeadEquation):

    def __init__(self, model, xc, yc, hc):
        Element.__init__(self, model, p=0)
        self.xc, self.yc = xc, yc
        self.pc = self.model.T * hc
        self.nunknowns = 1
```

```
    def omegainf(self, x, y):
        return np.ones_like(x, dtype='complex')
```

The example problem of Figure 8.2 (as discussed above) is solved with the analytic element code developed so far. The analytic element model is created with the following five commands:

```
# parameters and model
ml = Model(T=100) # transmissivity in m^2/d
w1 = Well(ml, xw=0, yw=0, Q=100, rw=0.3) # Q in m^3/d
w2 = HeadWell(ml, xw=400, yw=300, rw=0.3, hw=20)
uf = UniformFlow(ml, gradient=0.002, angle=0)
rf = Constant(ml, xc=400, yc=0, hc=22)
```

In the code cell above, the model is called ml. The unknown parameters in the model are computed by calling the solve method of the model in the code cell below. Next, the heads at the head-specified well and at the reference point are computed to verify that the solution is indeed correct. In addition, the head at the discharge-specified well and the computed discharge of the head-specified well are printed to the screen. Grids of the head and stream function are computed and the minimum and maximum values on the grid are printed to the screen, which are used to specify values for the contours to be plotted. Finally, a flow net is created (right graph in Figure 8.2). Note the (unfortunate) branch cuts in the stream function, which extend from each well to the left.

```
# solution
ml.solve() #compute the unknown parameters
print(f'computed head at discharge-specified well: {ml.head(0, 0):.2f} m')
print(f'computed head at head-specified well: {ml.head(400.3, 300):.2f} m')
print(f'computed discharge of head-specified well: {w2.parameter:.2f} m^3/d')
print(f'computed head at reference point: {ml.head(400, 0):.2f} m')
# grid
xg, yg = np.meshgrid(np.linspace(-200, 600, 100), np.linspace(-100, 500, 100))
head = ml.head(xg, yg)
psi = ml.streamfunction(xg, yg)
print(f'min and max head on grid: {head.min():.2f}, {head.max():.2f}')
print(f'min and max psi on grid: {psi.min():.2f}, {psi.max():.2f}')
```

```
computed head at discharge-specified well: 21.81 m
computed head at head-specified well: 20.00 m
computed discharge of head-specified well: 185.10 m^3/d
computed head at reference point: 22.00 m
min and max head on grid: 20.57, 23.47
min and max psi on grid: -137.29, 66.91
```

```
# basic flow net (right graph)
plt.subplot(111, aspect=1)
plt.contour(xg, yg, head, np.arange(20, 24, 0.2))
Q = w2.parameter # used to set stream function contour interval
plt.contour(xg, yg, psi, np.arange(-200, 100, Q / 20));
```

Exercise 8.2: Modify the HeadWell class to specify an optional location of the control point. Make a flow net for the previous example, but now the head for the head-specified well is 21.5 m and is specified at $(x, y) = (400, 200)$. Compute the head at the head-specified well and the discharge of the head-specified well.

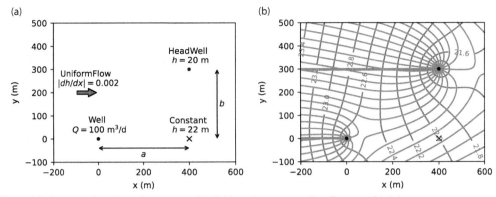

Figure 8.2 Layout of analytic element model (left) and corresponding flow net (right).

8.2   Line-sinks for modeling rivers and streams

Pumping wells are sometimes referred to as point-sinks because they take water out at a point (fully penetrating wells take water out over the entire aquifer thickness, which is a point in the two-dimensional horizontal plane). Line-sinks take out water along a line segment and may be used to simulate groundwater flow to or from sections of narrow rivers, streams, or canals. The analytic element code developed in the previous section is extended to include line-sinks. All analytic element classes `Model`, `Element`, `HeadEquation`, `Constant`, `UniformFlow`, `Well`, and `HeadWell`, defined in the previous section, are required to run the examples in this section.

The complex potential for a line-sink with uniform inflow may be obtained by integrating the complex potential for a point-sink (a well) along a line, which gives (e.g., Strack, 2017)

$$\Omega = \frac{\sigma L}{4\pi}[(Z+1)\ln(Z+1) - (Z-1)\ln(Z-1)] \tag{8.8}$$

where σ is the inflow into the line-sink (the discharge per unit length of line-sink, positive for taking water out of the aquifer), L is the length of the line-sink, and Z is defined as

$$Z = \frac{2\zeta - (\zeta_0 + \zeta_1)}{\zeta_1 - \zeta_0} \tag{8.9}$$

where ζ is the complex coordinate, and ζ_0 and ζ_1 are the complex coordinates of the end points of the line-sink. Note that $Z = -1$ at $\zeta = \zeta_0$ and $Z = +1$ at $\zeta = \zeta_1$. The total discharge Q of the line-sink is $Q = \sigma L$.

Line-sinks are added to the analytic element code for wells and uniform flow developed in Section 8.1. Input arguments are the model to which the element is added, the coordinates of the endpoints, and the inflow σ. The `LineSink` class has one function: the `omegainf` function, which implements Eq. (8.8) for $\sigma = 1$. The complex potential for a line-sink is singular at the end points, where the variable Z equals either -1 or $+1$, but the discharge potential is finite

$$\lim_{Z \to -1}(Z+1)\ln(Z+1) = \lim_{Z \to 1}(Z-1)\ln(Z-1) = 0 \tag{8.10}$$

The value of $(Z-1)$ or $(Z+1)$ is set to 10^{-12} when their absolute value is smaller than 10^{-12} using the `np.where` function in the code below.

```
class LineSink(Element):

    def __init__(self, model, x0=0, y0=0, x1=1, y1=1, sigma=1):
        Element.__init__(self, model, sigma)
        self.z0 = x0 + y0 * 1j
        self.z1 = x1 + y1 * 1j
        self.L = np.abs(self.z1 - self.z0)

    def omegainf(self, x, y):
        zeta = x + y * 1j
        Z = (2 * zeta - (self.z0 + self.z1)) / (self.z1 - self.z0)
        Zp1 = np.where(np.abs(Z + 1) < 1e-12, 1e-12, Z + 1)
        Zm1 = np.where(np.abs(Z - 1) < 1e-12, 1e-12, Z - 1)
        return self.L / (4 * np.pi) * (Zp1 * np.log(Zp1) - Zm1 * np.log(Zm1))
```

As a first example, the flow net for a single line-sink is computed and plotted in the left graph of Figure 8.3. The branch cut starts at the first point of the line-sink and extends to infinity in the same direction as the line-sink. The variation of the head along the line-sink is plotted in the right graph of Figure 8.3.

```
# parameters and model
ml = Model(T=100) # transmissivity, m^2/d
ls1 = LineSink(ml, x0=-200, y0=-150, x1=200, y1=150, sigma=0.1) # sigma, m^2/d
```

```
# solution
xg, yg = np.meshgrid(np.linspace(-400, 400, 100), np.linspace(-300, 300, 100))
h = ml.head(xg, yg)
psi = ml.streamfunction(xg, yg)
xs, ys = np.linspace(-200, 200, 100), np.linspace(-150, 150, 100)
hs = ml.head(xs, ys) # head along line-sink
```

```
# basic plot
plt.subplot(121, aspect=1)
plt.contour(xg, yg, h, 10)
plt.contour(xg, yg, psi, 20)
plt.plot([-200, 200], [-150, 150], 'k')
plt.subplot(122)
plt.plot(np.sqrt((xs + 200) **2 + (ys + 150) ** 2), hs);
```

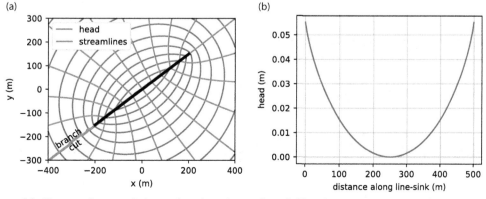

Figure 8.3 Flow net for a single line-sink with uniform inflow (left) and head along line-sink (right).

The inflow is uniform along a line-sink, which results in a head that varies along the line-sink (right graph of Figure 8.3). This means that a finite ditch with a uniform head cannot be modeled with a single line-sink of uniform inflow. Such a ditch must be simulated with a string of multiple line-sinks, as the inflow near the ends of the ditch is larger than in the middle of the ditch.

The free parameter of a line-sink is the inflow σ into the line-sink per unit length of line-sink. When a line-sink is used to simulate the segment of a surface water feature, the inflow (negative when it is outflow) may be computed from a specified-head condition. Here, the head is specified at the center of the line-sink (the collocation point). In the code below, the HeadLineSink class is derived from the LineSink and HeadEquation classes. Input arguments are the model to which the element is added, coordinates of the endpoints, and the specified head at the center of the line-sink. As for all head-specified elements, the location of the collocation point is stored as xc and yc, and the potential at the collocation point as pc. The number of unknowns is set to 1.

```
class HeadLineSink(LineSink, HeadEquation):

    def __init__(self, model, x0=0, y0=0, x1=1, y1=1, hc=1):
        LineSink.__init__(self, model, x0, y0, x1, y1, 0)
        self.xc = 0.5 * (x0 + x1)
        self.yc = 0.5 * (y0 + y1)
        self.pc = self.model.T * hc
        self.nunknowns = 1
```

Strings of line-sinks may be used to simulate narrow rivers or streams. A stream needs to be subdivided into line-sinks based on three criteria: The string of line-sinks must represent the geometry of the stream, and the line-sinks must be short enough to simulate both the variation in river stage and the variation of the inflow or outflow along the stream. Recall that the inflow is uniform for each line-sink. In the current implementation of head-specified line-sinks, the head in the aquifer is set equal to the specified head at the center of the line-sink, which means that the aquifer is in direct contact with the stream. Alternatively, the resistance of a leaky stream bed may be taken into account by specifying condition (7.34) of Section 7.4, but that is beyond the scope of this book (see, e.g., Haitjema, 1995).

As an example of a curved stream, consider a well pumping near a meandering stream. The stream is approximated with a sine function around the x-axis as $y = 50 \sin(\pi x/400)$. A head-specified well is located at $(x, y) = (0, 100)$ with a head that is 2 m below the head in the river. The head is fixed to 2 m above the river level at $(x, y) = (0, 1000)$. A stretch of stream is modeled from $x = -1600$ m to $x = 1600$ m, which is far enough to the left and right of the well such that a further extension does not affect the heads in the area around the well significantly. Shorter line-sinks with a length of 50 m are used to represent the stream near the well, while longer line-sinks with a length of 200 m are used farther away. All other parameters are given in the script below. A contour plot of the head for the entire model is shown in the left graph of Figure 8.4, while a contour plot of the area around the well is shown in the right graph of Figure 8.4.

```
# parameters and model
hriver = 10 # head in river, m
ml = Model(T=100) # transmissivity, m^2/d
rf = Constant(ml, xc=0, yc=1000, hc=hriver + 2)
w = HeadWell(ml, xw=0, yw=100, rw=0.3, hw=hriver - 2)
xls = np.linspace(-1600, 1600, 101) # x-locations of line-sinks, m
xls = np.hstack((np.arange(-1600, -400, 200),
```

```
       np.arange(-400, 400, 50), np.arange(400, 1601, 200)))
yls = 50 * np.sin(np.pi * xls / 400) # y-locations of line-sinks, m
for i in range(len(xls) - 1):
    HeadLineSink(ml, x0=xls[i], y0=yls[i],
                    x1=xls[i + 1], y1=yls[i + 1], hc=hriver)
```

```
# solution
ml.solve()
xg1, yg1 = np.meshgrid(np.linspace(-1800, 1800, 100),
                       np.linspace(-1200, 1200, 100))
h1 = ml.head(xg1, yg1) # entire model
xg2, yg2 = np.meshgrid(np.linspace(-400, 400, 100), np.linspace(-100, 400, 100))
h2 = ml.head(xg2, yg2) # model around well
```

```
# basic contour plot
plt.subplot(121, aspect=1)
plt.contour(xg1, yg1, h1, 10)
plt.plot(xls, yls, 'k')
plt.subplot(122, aspect=1, xlim=(-400, 400))
plt.contour(xg2, yg2, h2, 10)
plt.plot(xls, yls, 'k');
```

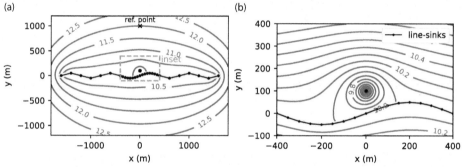

Figure 8.4 Head contours for a well near a meandering river; entire model (left) and contours near the well (right) for the green inset in the left plot.

Exercise 8.3: Compare a line-sink to 10 evenly spaced wells. Draw head contours for one line-sink of 200 m length and $\sigma = 2$ m^2/d. On the same plot, draw head contours for 10 evenly spaced wells with a total discharge equal to the total discharge of the line-sink. The transmissivity of the aquifer is $T = 100$ m^2/d, and the endpoints of the line-sink are at $(-100, 0)$ and $(100, 0)$. Specify the head to 20 m far away at $(0, 10000)$.

Exercise 8.4: Simulate flow to a ditch with specified head of 10 m. The ditch runs from $(-200, -100)$ till $(200, 100)$. The head at $(0, 400)$ is 12 m. Use enough line-sinks such that the head along the ditch is approximately uniform. The transmissivity is 200 m^2/d. Create a flow net and make separate plots of the head variation along the ditch and the inflow variation along the ditch. Compute the total inflow into the ditch.

Exercise 8.5: Consider a well pumping near a long and straight river along the x-axis. The transmissivity of the aquifer is 200 m^2/d, and the head in the river is 8 m. The well is located at $(0, 100)$, the well radius is 0.3 m, and the head in the well is 6 m. Simulate the river with head-specified line-sinks. Use enough line-sink of an appropriate length such that the analytic element solution is in reasonable agreement with the exact solution obtained with the method of images (Exercise 6.8). Compute the discharge of the well.

8.3   Area-sinks for modeling areal recharge

Bounded areas with a specified recharge are called area-sinks. An area-sink is the final type of analytic element that is added to the analytic element code developed in this chapter. All analytic element classes Model, Element, HeadEquation, Constant, UniformFlow, Well, HeadWell, LineSink, and HeadLineSink, defined in the previous sections, are required to run the examples in this section.

In this section, circular area-sinks are discussed with a uniform recharge. Rectangular area-sinks or area-sinks bounded by an arbitrary closed polygon also exist (e.g., Strack, 2017), but their equations are significantly more complicated and beyond the scope of this book. Circular area-sinks are very useful for the simulation of areal recharge from precipitation. Their circular geometry is commonly not a problem as long as the area-sink covers the entire model area, and heads are fixed inside the model area by, e.g., head-specified line-sinks, as shown in the example at the end of this section.

Consider a circular area-sink of radius R centered at $(x, y) = (x_a, y_a)$. Inside the area-sink, the recharge is uniform and equal to N. Note that N is positive for water entering the aquifer (this is opposite to wells and line-sinks where the discharge Q and discharge per unit length σ are positive for water leaving the aquifer). The governing differential equation inside the area-sink is

$$\nabla^2 \Phi = \frac{d^2\Phi}{dr^2} + \frac{1}{r}\frac{d\Phi}{dr} = -N \qquad r \leq R \tag{8.11}$$

where $r = \sqrt{(x - x_a)^2 + (y - y_a)^2}$ is a local radial coordinate. The areal recharge equals zero outside the area-sink, where the governing differential equation is the Laplace equation (Eq. 6.11). The solution inside the area-sink was already derived (Eq. 6.29) as

$$\Phi = -\frac{1}{4}N(r^2 - R^2) \qquad r \leq R \tag{8.12}$$

Outside the area-sink, the solution is equal to the solution for an injection well with injection rate $N\pi R^2$ and zero potential at $r = R$

$$\Phi = -\frac{NR^2}{2}\ln(r/R) \qquad r \geq R \tag{8.13}$$

Note that the potential (and hence the head) and the discharge vector are continuous across the edge of the area-sink. The stream function doesn't exist inside the area-sink, as the stream function exists only when the areal recharge equals zero, and is not used outside the area-sink.

The AreaSink class is defined below. Input arguments are the model that the area-sink is added to, the location of the center of the area-sink, the uniform recharge inside the area-sink, and the radius of the area-sink. The AreaSink class stores all input arguments and includes the omegainf function, where the real part is the potential defined above (Eqs. (8.12) and (8.13) with $N = 1$) and the imaginary part equals zero.

```
class AreaSink(Element):

    def __init__(self, model, xc=0, yc=0, N=0.001, R=100):
        Element.__init__(self, model, N)
        self.xc, self.yc = xc, yc
        self.R = R

    def omegainf(self, x, y):
        r = np.atleast_1d(np.sqrt((x - self.xc) ** 2 + (y - self.yc) ** 2))
```

```
phi = np.zeros(r.shape)
phi[r < self.R] = -0.25 * (r[r < self.R] ** 2 - self.R ** 2)
phi[r >= self.R] = -self.R ** 2 / 2 * np.log(r[r >= self.R] / self.R)
return phi + 0j
```

As a first example, consider the synthetic case of two area-sinks next to each other. One is a recharge area, and the other is a discharge area. The total amount of recharge is equal to the total amount of discharge so that the flow field is symmetric and the problem can be used to verify the implementation of the solution. Contours of the head are shown on the left side of Figure 8.5, and a cross section through the centers of the two area-sinks is shown on the right side. Note that the highest and lowest heads are not at the centers of the area-sinks.

```
# parameters and model
ml = Model(T=100) # transmissivity, m^2/d
rf = Constant(ml, xc=0, yc=0, hc=20)
as1 = AreaSink(ml, xc=-500, yc=0, N=0.001, R=500) # recharge, m/d
as2 = AreaSink(ml, xc=500, yc=0, N=-0.001, R=500) # discharge, m/d
```

```
# solution
ml.solve()
xg, yg = np.meshgrid(np.linspace(-1500, 1500, 100), np.linspace(-800, 800, 101))
h = ml.head(xg, yg)
```

```
# basic plot
plt.subplot(121, aspect=1)
plt.contour(xg, yg, h, 20)
plt.subplot(122)
plt.plot(xg[0], h[50]);
```

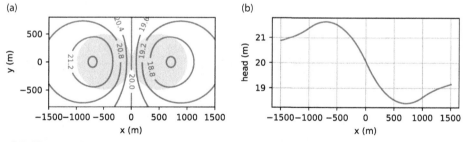

Figure 8.5 Two circular area-sinks; head contours (left) and head along x-axis (right).

Exercise 8.6: Replicate the final example of Section 6.1 using analytic elements for the case that the well pumps half the recharge on the island. Approximate the transmissivity as $k(h_R - z_b)$. Plot the analytic element solution with constant transmissivity and the exact solution for unconfined flow of Section 6.1 on the same graph.

As a final example, all analytic elements are combined in one model with two pumping wells located between two river branches. Each river branch is simulated with 8 line-sinks. The locations and discharges of the wells, and the coordinates and heads at the centers of the river segments, are given below and plotted in Figure 8.6 (left graph). Areal recharge is added with a circular area-sink that is large enough to cover the area between the rivers.

The head is specified at a point outside the model area; such a point is called the reference point in analytic element models. The specified head at the reference point controls how much water flows from infinity toward the model area. A lot of water flows toward the model area when the head at the reference point is high, while water flows away from the model area toward infinity when the head at the reference point is low. As such, the reference point is a model feature and not a physical feature. The head at the reference point does not affect the head in the model area significantly if the model area is surrounded by enough head-specified elements (e.g., rivers, streams, and canals). It is recommended to specify a head at the reference point such that the head gradients outside the modeled area are similar to the head gradients inside the model area.

```
# river segments, all in m
xls0 = [0, 100, 200, 400, 600, 800, 1000, 1100, 1200]
yls0 = [200, 200, 100, 100, 0, 0, 100, 300, 450]
hls0 = np.linspace(39, 40.4, 8)
xls1 = [0, 0, 200, 400, 600, 800, 1000, 1100, 1200]
yls1 = [200, 400, 600, 600, 700, 700, 750, 800, 850]
hls1 = np.linspace(39, 40.4, 8)
```

```
# parameters and model
ml = Model(T=100) # transmissivity, m^2/d
rf = Constant(ml, xc=0, yc=800, hc=39.5)
w0 = Well(ml, xw=500, yw=250, Q=100) # discharge, m^3/d
w1 = Well(ml, xw=800, yw=500, Q=100) # discharge, m^3/d
for i in range(len(hls0)):
    HeadLineSink(ml, xls0[i], yls0[i], xls0[i + 1], yls0[i + 1], hls0[i])
for i in range(len(hls1)):
    HeadLineSink(ml, xls1[i], yls1[i], xls1[i + 1], yls1[i + 1], hls1[i])
ar = AreaSink(ml, xc=600, yc=400, N=0.001, R=700) # recharge, m/d
```

```
# solution
ml.solve()
xg, yg = np.meshgrid(np.linspace(-100, 1300, 100), np.linspace(-100, 900, 100))
h = ml.head(xg, yg)
```

```
# basic contour plot (right side of figure)
plt.subplot(111, aspect=1)
cs = plt.contour(xg, yg, h, np.arange(38, 41, 0.2), colors='C0')
plt.clabel(cs, fmt='%1.1f', fontsize='smaller')
plt.contour(xg, yg, h, np.arange(38.1, 41, 0.2), colors='C0')
plt.plot(xls0, yls0, 'k')
plt.plot(xls1, yls1, 'k')
plt.plot([w0.zw.real, w1.zw.real], [w0.zw.imag, w1.zw.imag], 'k.');
```

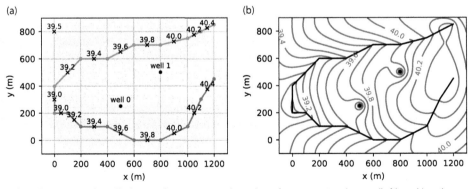

Figure 8.6 Example with wells, line-sinks, an area-sink, and a reference point; layout (left) and head contours (right).

Chapter 9

Transient two-dimensional flow

All two-dimensional solutions presented so far were for steady flow: The head and flow are a function of x and y, but not of time t. In this chapter, wells are considered with a discharge that varies through time. Both the head and flow are a function of x, y, and t. The changes in head and flow caused by transient wells are discussed for a variety of settings.

Continuity of flow for two-dimensional transient flow in the horizontal x, y plane may be written as

$$\frac{\partial Q_x}{\partial x} + \frac{\partial Q_y}{\partial y} = -S\frac{\partial h}{\partial t} + N \tag{9.1}$$

where N is the areal recharge and S is the storage coefficient, which is equal to the phreatic storage $S = S_p$ for unconfined flow and the product of the specific storage S_s and aquifer thickness H ($S = S_s H$) for confined flow (see Chapter 5). Note that this equation differs from the continuity of flow equation for one-dimensional transient flow (Eq. 5.5) only by the additional term $\partial Q_y/\partial y$.

The components of the discharge vector are written in terms of the derivatives of the head as

$$Q_x = -T\frac{\partial h}{\partial x} \qquad Q_y = -T\frac{\partial h}{\partial y} \tag{9.2}$$

where the transmissivity is approximated as constant, also for unconfined flow (see Chapter 5). Substitution of Eq. (9.2) for Q_x and Q_y in Eq. (9.1) gives the differential equation for two-dimensional transient flow

$$\nabla^2 h = \frac{S}{T}\frac{\partial h}{\partial t} - \frac{N}{T} \tag{9.3}$$

The discharge potential is not very useful for transient flow (except for the alternative formulation for unconfined flow discussed in Section 5.5), and the stream function and complex potential are only valid for flow governed by the Laplace equation, so they are not used here. For the case that areal recharge is negligible ($N = 0$), Eq. (9.3) reduces to the two-dimensional diffusion equation

$$\nabla^2 h = \frac{S}{T}\frac{\partial h}{\partial t} \tag{9.4}$$

If, in addition, the flow is radially symmetric, the differential equation becomes (compare Eq. (6.8))

$$\frac{\partial^2 h}{\partial r^2} + \frac{1}{r}\frac{\partial h}{\partial r} = \frac{S}{T}\frac{\partial h}{\partial t} \tag{9.5}$$

DOI: 10.1201/9781315206134-9

Solutions to the two-dimensional diffusion equation are difficult to derive mathematically, even for many cases of radially symmetric flow. Mathematical solution techniques to obtain exact solutions include similarity solutions, separation of variables, and a variety of transform techniques (Laplace, Hankel, and Fourier). In the following, several solutions for transient flow to wells are presented rather than derived. The solutions are discussed, programmed, and visualized as in the other chapters. The mathematical correctness of the presented solutions may be verified through differentiation of the solution and substitution into the differential equation (either analytically or numerically), and by verifying that the boundary conditions and initial conditions are met.

All solutions presented in this chapter are for the head change caused by a transient well, which allows for the superposition of transient solutions in time, and for the superposition of steady solutions and transient solutions. Transient flow solutions are presented for wells in confined, unconfined, semi-confined, and two-aquifer systems, including the estimation of aquifer parameters from pumping tests. The last two sections concern the effect of wellbore storage and skin effect, and transient pumping in a two-aquifer system. These problems are solved in the Laplace domain, and the solution in the physical domain is obtained through a numerical back-transformation.

9.1   Wells in confined and unconfined aquifers

In this section, a solution is presented for transient flow to a well in an aquifer with constant transmissivity T and storage coefficient S. The governing differential equation is (9.5). The well starts pumping with discharge Q at time $t = t_0$. The boundary condition at the well is approximated as (similar to boundary condition (6.24))

$$\lim_{r \to 0} -2\pi r Q_r = Q \qquad t \geq t_0 \tag{9.6}$$

Initially, the head change in the aquifer caused by the transient well is, of course, equal to zero

$$h|_{r, t=t_0} = 0 \tag{9.7}$$

The second boundary condition is that the head change caused by the well at infinity is zero at all time

$$h|_{r=\infty, t} = 0 \tag{9.8}$$

The solution for the head is a function of the radial distance, r, from the well and time and is known as the Theis solution (Theis, 1935)

$$h = -\frac{Q}{4\pi T} E_1(u) \qquad t \geq t_0 \tag{9.9}$$

where

$$u = \frac{Sr^2}{4T(t - t_0)} \tag{9.10}$$

and E_1 is the exponential integral defined as

$$E_1(u) = \int_u^\infty \frac{\exp(-s)}{s} ds \tag{9.11}$$

The head starts to decrease after the well starts pumping. The drop in head is also referred to as the drawdown. So a drawdown of, e.g., 1 m means that $h = -1$ m. The head is a function only of the dimensionless parameter u. Hence, if a certain drawdown is reached at a distance r_1 after a time period Δt_1 of pumping, the same drawdown is reached at a distance $2r_1$ after a time period $4\Delta t_1$ of pumping, as both combinations of time and radial distance result in the same value of u

$$u = \frac{Sr_1^2}{4T\Delta t_1} = \frac{S(2r_1)^2}{4T(4\Delta t_1)} \tag{9.12}$$

Exercise 9.1: A well starts pumping at time $t = 0$ in a confined aquifer. After 1 day of pumping, the drawdown is 1 m at a distance of $r_1 = 20$ m from the well. Compute at what time (in days) after pumping started the drawdown is 1 m at a distance of $r_2 = 60$ m from the well.

Exercise 9.2: Consider a well with discharge $Q = 300$ m^3/d that is screened in a confined aquifer with transmissivity $T = 500$ m^2/d and storage coefficient $S = 0.2$. At what distance from the well is the drawdown 0.2 m after 1 day of pumping? Hint: use the `fsolve` function of `scipy.optimize`.

Exercise 9.3: Consider a well with discharge $Q = 300$ m^3/d and radius $r_w = 0.3$ m. The well has been pumping in an aquifer with $T = 500$ m^2/d and $S = 0.2$ for 1 day. Compute the water removed from storage by evaluating the integral $\int_{r_w}^{\infty} Sh2\pi r dr$ using a numerical integration function from `scipy.integrate`. Compare your answer to the total amount of water removed from the aquifer by the well.

The function E_1 is available as `exp1` from the `scipy.special` package, but there also exists a series representation, which is useful for the evaluation of limiting cases (e.g., large time). The series representation is (e.g., Abramowitz and Stegun, 1965)

$$E_1(u) = -\gamma - \ln u - \sum_{n=1}^{\infty} \frac{(-u)^n}{n \cdot n!} \tag{9.13}$$

where $\gamma = 0.5772...$ is Euler's constant, available from `numpy` as `euler_gamma`. The series for E_1 converges quickly when $u < 1$. Only the first two terms are needed when u is small, in which case the formula for the head may be approximated as

$$h \approx \frac{Q}{4\pi T}\left[\gamma + \ln\left(\frac{Sr^2}{4T(t - t_0)}\right)\right] \qquad u < 0.1 \tag{9.14}$$

Note that u is small when r is small or when t is large, or both.

The head inside the well is computed by evaluating the Theis solution at the well radius r_w. Four curves are shown in the graphs of Figure 9.1: the head h_w inside the well, the head h_{20} at $r = 20$ m, the difference $h_w - h_{20}$, and the head h_{20} computed with the approximate formula (9.14). The time axis is linear in the left graph and log-scaled in the right graph. The difference $h_w - h_{20}$ becomes constant after ~ 0.01 d for this case. The approximate formula results in a straight line in the graph with a log-scaled time axis, as can also be seen from Eq. (9.14). The approximate formula gives good results for $u < 0.1$, which corresponds to $t > 0.01$ d for this case and at this location.

```
# parameters
T = 200 # transmissivity, m^2/d
S = 2E-3 # storage coefficient, -
rw = 0.3 # radius of well, m
```

```
Q = 1000 # discharge of well, m^3/d
t0 = 0 # start of pumping, d
```

```
# solution
from scipy.special import exp1
def htheis(r, t, T, S, Q, t0=0):
    return -Q / (4 * np.pi * T) * exp1(S * r ** 2 / (4 * T * (t - t0)))

t = np.logspace(-3, 0, 100)
hw = htheis(rw, t, T, S, Q, t0)
h20 = htheis(20, t, T, S, Q, t0)
h20approx = Q / (4 * np.pi * T) * (np.euler_gamma +
                            np.log(S * 20 ** 2 / (4 * T * (t - t0))))
```

```
# basic plot
plt.subplot(121)
plt.plot(t, hw, label='$h_w$')
plt.plot(t, h20, label='$h_{20}$')
plt.plot(t, h20approx, '--', label='$h_{20}$ approx')
plt.plot(t, hw - h20, label='$h_w - h_{20}$')
plt.legend()
plt.subplot(122)
plt.semilogx(t, hw)
plt.semilogx(t, h20)
plt.semilogx(t, h20approx, '--')
plt.semilogx(t, hw - h20);
```

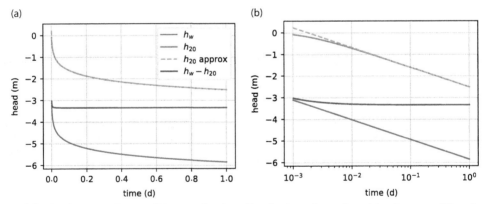

Figure 9.1 Head vs. time for a well in a confined aquifer; h_w is at the well, and h_{20} is at $r = 20$ m; linear time axis (left) and log-scaled time axis (right).

It may be expected that the head approaches steady state when a well is pumped for a long enough period of time. Strangely enough, this is not the case: The head for a transient well (Eq. 9.9) does not approach the head for a steady well in an infinite aquifer (Eq. 6.37) for large time. For a steady well (Eq. 6.37), the head approaches infinity when r approaches infinity, because the source of water for a steady well in an infinite aquifer lies at infinity. For a transient well, the head approaches zero when r approaches infinity. This means that all the water pumped by a transient well in the Theis solution comes from storage. For a single transient well pumping in an infinite aquifer, it is not possible to determine the distance from the well where the drawdown equals zero, because theoretically the drawdown only equals zero at infinity. The drawdown is, obviously, already negligible at a much shorter distance from the well. The distance from the well where the drawdown is negligible can be estimated by rewriting the approximate formula for large time (Eq. 9.14) in the form of the Thiem equation (Eq. 6.25)

$$h \approx \frac{Q}{2\pi T} \ln(r/R) \qquad r \leq R \tag{9.15}$$

where the distance R is a function of time

$$R = \sqrt{4T(t - t_0)e^{-\gamma}/S} \tag{9.16}$$

The distance R may be used as an approximation of the distance at which the drawdown may be neglected.

Exercise 9.4: Plot the head vs. r at time $t = 1$ d for r varying from r_w to 600 m using the parameters of the example. On the same graph, plot the approximate formula (9.15).

Even though the head of the Theis solution by itself does not approach the steady-state head of a steady well, the difference between the head at two locations does approach a constant value. This was already shown in Figure 9.1, where the difference in the head between $r = r_w$ and $r = 20$ m approaches a constant value for $t > 0.01$ d ($u < 0.1$) for this case. Similarly, the discharge vector approaches the steady solution for small u. The radial component of the discharge vector Q_r of the Theis solution may be obtained from Eq. (9.9) as

$$Q_r = -T\frac{\partial h}{\partial r} = -T\frac{dh}{du}\frac{\partial u}{\partial r} = -\frac{Q}{2\pi}\frac{1}{r}e^{-u} \tag{9.17}$$

When time is large, u approaches zero and Q_r approaches the steady discharge vector

$$Q_r = -\frac{Q}{2\pi}\frac{1}{r} \tag{9.18}$$

Exercise 9.5: Compute the steady head difference between $r = r_w$ and $r = 20$ m, and compare your answer to the difference obtained with the transient solution for time varying from $t = 0.01$ to $t = 1$ d.

In practice, there is always a water source closer than infinity, and if that source is included in the solution, the transient solution approaches a steady solution for large time. For example, consider a well at $(x, y) = (-d, 0)$ near a large and straight river with a constant head h_0 along $x = 0$. The steady solution may be obtained with the method of images (Section 6.2). A transient solution may also be obtained with the method of images as

$$h = -\frac{Q}{4\pi T}\left[E_1\left(\frac{Sr_1^2}{4T(t - t_0)}\right) - E_1\left(\frac{Sr_2^2}{4T(t - t_0)}\right)\right] + h_0 \tag{9.19}$$

where $r_1^2 = (x + d)^2 + y^2$ and $r_2^2 = (x + d)^2 + y^2$. For large time, E_1 may again be represented with the first two terms of the series representation (Eq. 9.13). Substitution of these terms for E_1 into Eq. (9.19) and some algebra leads to the steady solution

$$h = \frac{Q}{2\pi T}\ln\frac{r_1}{r_2} + h_0 \tag{9.20}$$

Transient solutions may be superimposed in time as well as in space. For example, consider a well with a discharge Q operating from $t = t_0$ to $t = t_1$ and with zero discharge after t_1. For the period $t > t_1$, the head function consists of two Theis wells, one with discharge Q starting at $t = t_0$ and one with discharge $-Q$ starting at $t = t_1$

$$h = -\frac{Q}{4\pi T}\left[E_1\left(\frac{Sr^2}{4T(t - t_0)}\right) - E_1\left(\frac{Sr^2}{4T(t - t_1)}\right)\right] \qquad t \geq t_1 \tag{9.21}$$

The variation of the head after the well is turned off is called the recovery phase, as the head (slowly) comes back to its original position. For large time, E_1 may again be represented with the first two terms of the series representation of E_1 (Eq. 9.13), which results in

$$h \approx -\frac{Q}{4\pi T} \ln\left(\frac{t - t_1}{t - t_0}\right) \qquad t \gg t_1 \tag{9.22}$$

This is an important and somewhat surprising result, as the recovery after the well has been turned off for a little while does not depend on the storage coefficient S. In the example below, two solutions are plotted for a well that pumps for one day and is then turned off, but the storage coefficients differ. The head response is quite different during pumping, but the two curves quickly approach each other during the recovery phase, even though the storage coefficients are different. The approximate formula (9.22) is also plotted for the recovery phase in Figure 9.2.

```
# parameters
t0 = 0 # time pump starts, d
t1 = 1 # time pump is turned off, d
S = [1E-3, 1E-4] # storage coefficients, -
r = 100 # m
```

```
# solution
t = np.linspace(1e-6, 2, 200)
h = np.zeros((2, len(t)))
for i in range(2):
    h[i] = -Q / (4 * np.pi * T) * exp1(S[i] * r ** 2 / (4 * T * (t - t0)))
    h[i, t > t1] -= -Q / (4 * np.pi * T) * exp1(
        S[i] * r ** 2 / (4 * T * (t[t > t1] - t1)))
tapprox = np.linspace(t1 + 0.01, 2, 100)
happrox = Q / (4 * np.pi * T) * np.log((tapprox - t1) / (tapprox - t0))
```

```
# basic plot
plt.plot(t, h[0])
plt.plot(t, h[1])
plt.plot(tapprox, happrox);
```

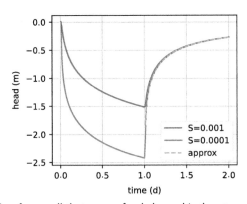

Figure 9.2 Head at $r = 100$ m for a well that pumps for 1 day and is then turned off; results are shown for two different storage coefficients, and for the approximate solution during the recovery phase (Eq. 9.22).

The Theis solution is very useful to estimate aquifer parameters from a pumping test, also called an aquifer test. During a pumping test, a well is turned on and the head (drawdown) is measured in one or more nearby observation wells and/or the pumping well itself. The Theis solution is fitted to the observed heads to estimate the transmissivity T and the storage coefficient S of the aquifer.

Kruseman and De Ridder (1990) described a pumping test at the *Oude Korendijk* in The Netherlands in their famous book on pumping tests. The objective of the pumping test was to determine the transmissivity and storage coefficient of the aquifer. The discharge of the well during the pumping test was 788 m³/day. The observed drawdown (in meters) and the observation time (in minutes) after pumping started are given below for an observation well located 30 m from the pumping well. Kruseman and De Ridder (1990) also reported drawdown in two additional wells, but only one observation well is used here to demonstrate the approach. Time is converted to days, and drawdown is converted to heads.

```
# Data at Oude Korendijk taken from Kruseman and de Ridder (1990)
robs = 30 # distance to observation well, m
Q = 788 # discharge of well, m^3/d
time = np.array(
    [0.1 ,  0.25,  0.5 ,  0.7 ,  1.  ,  1.4 ,  1.9 ,  2.33,  2.8 ,
     3.36,  4.  ,  5.35,  6.8 ,  8.3 ,  8.7 , 10.  , 13.1 , 18.  ,
    27.  , 33.  , 41.  , 48.  , 59.  , 80.  , 95.  , 139. , 181. ,
    245., 300., 360., 480., 600., 728., 830.]) # in minutes
drawdown = np.array(
    [0.04 , 0.08 , 0.13 , 0.18 , 0.23 , 0.28 , 0.33 , 0.36 , 0.39 ,
     0.42 , 0.45 , 0.5  , 0.54 , 0.57 , 0.58 , 0.6  , 0.64 , 0.68 ,
     0.742, 0.753, 0.779, 0.793, 0.819, 0.855, 0.873, 0.915, 0.935,
     0.966, 0.99 , 1.007, 1.05 , 1.053, 1.072, 1.088]) # in meters
tobs = time / 24 / 60 # convert observation time to days
hobs = -drawdown # convert drawdown to heads
```

The values of the transmissivity and the storage coefficient are estimated using a least squares approach by minimizing the sum of squared differences between the modeled heads and the observed heads (similar to the approach in Section 5.3). The objective function fobj returns the sum of squared errors and takes five input arguments: an array of the parameters T and S, the observed heads, observation times, distance to observation well, and discharge of the well. When the keyword argument return_heads is set to True, the function fobj returns the modeled heads rather than the sum of squared errors.

```
def fobj(p, ho, to, ro, Q, return_heads=False):
    T, S = p
    hm = htheis(ro, to, T, S, Q)
    if return_heads:
        return hm
    rv = np.sum((ho - hm) ** 2)
    return rv
```

The fmin function of the scipy.optimize package is used to find the parameters that minimize the objective function. The four arguments ho, to, ro, and Q of fobj are passed by fmin to the objective function fobj with the args keyword argument of fmin. The fmin function does not estimate the uncertainty of the estimated parameters, but that is not a problem here as standard procedures to estimate the uncertainty are commonly based on certain statistical requirements between the modeled and observed heads that are violated in many pumping test analyses (e.g., the differences between the observed and modeled heads are not independent). The observed heads and fitted curve are plotted on the same graph in Figure 9.3. The left plot has a linear time axis. The right plot has a semi-log axis, which makes it easier to see the head variation at early time.

```
# least squares solution
from scipy.optimize import fmin
T, S = fmin(fobj, [100, 1e-4], args=(hobs, tobs, robs, Q), disp=0)
print(f'Estimated parameters T: {T:.2f} m^2/d, S: {S:.2e}')
hm = fobj([T, S], hobs, tobs, robs, Q, return_heads=1) # compute modeled heads
rmse = np.sqrt(np.mean((hobs - hm) ** 2))
print(f'Root mean squared error: {rmse:.3f} m')
```

```
Estimated parameters T: 480.47 m^2/d, S: 1.13e-04
Root mean squared error: 0.032 m
```

```
# basic plot
plt.subplot(121)
plt.plot(tobs, hobs, 'C1.')
plt.plot(tobs, hm, 'C0')
plt.subplot(122)
plt.semilogx(tobs, hobs, 'C1.')
plt.semilogx(tobs, hm, 'C0');
```

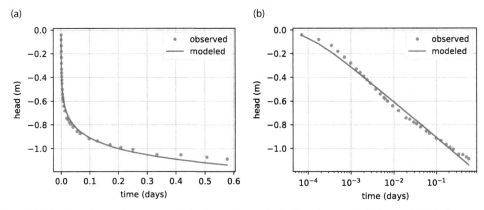

Figure 9.3 Results of pumping test analysis; linear time axis (left) and log-scaled time axis (right).

Exercise 9.6: Consider a pumping test for a well that is located 30 m from a river. Measurements are taken in an observation well located 5 m from the well and 25 m from the river. Hourly measurements of a 20-hour pumping test are given below. The discharge of the well was 200 m^3/d. Estimate the transmissivity and storage coefficient of the aquifer from the pumping test first by neglecting the presence of the river and second by including the river using the method of images.

```
tobs = np.array([1, 2, 3, 4, 5, 6, 7, 8, 9, 10,       # time in hours
    11, 12, 13, 14, 15, 16, 17, 18, 19, 20])
hobs = np.array(                                        # head change in m
    [-0.16, -0.22, -0.24, -0.27, -0.28, -0.29, -0.31, -0.31, -0.32, -0.32,
    -0.33, -0.33, -0.34, -0.34, -0.35, -0.35, -0.34, -0.34, -0.35, -0.35])
```

9.2 Wells with a periodic discharge

The discharge of a well may vary through time periodically. For example, in an aquifer storage and recovery (ASR) system, water is injected into the aquifer during times of abundant water availability and pumped out of the aquifer during times of water scarcity. For an ASR well, the discharge varies around zero, for example sinusoidally as

$$Q(t) = A \sin(2\pi t/\tau) \tag{9.23}$$

where A and τ are the amplitude and period of the sinusoidal discharge. The boundary condition at the well is approximated as

$$\lim_{r \to 0} -2\pi r Q_r = A \sin(2\pi t/\tau) \tag{9.24}$$

while the boundary condition at infinity is (9.8). There is no initial condition: The well has been pumping with the sinusoidal discharge (Eq. 9.23) for a long time. Flow in the aquifer is governed by the diffusion equation (9.5).

The solution for the stated problem is obtained using separation of variables: The solution is written as a function of the radial distance r multiplied by a function of the time t. The analysis is easiest conducted in terms of complex variables. The head solution corresponding to boundary condition (9.23) is the imaginary part of the complex solution (Bruggeman, 1999)

$$h = -\frac{A}{2\pi T} \Im \left[\mathrm{K}_0 \left(\frac{r}{\mu} \sqrt{i} \right) e^{2\pi it/\tau} \right] \tag{9.25}$$

where the characteristic length μ is defined as

$$\mu = \sqrt{\frac{T}{S} \frac{\tau}{2\pi}} \tag{9.26}$$

Note that μ has the dimension of length. The corresponding radial component of the discharge vector is

$$Q_r = -\frac{A}{2\pi} \Im \left[\frac{\sqrt{i}}{\mu} \mathrm{K}_1 \left(\frac{r}{\mu} \sqrt{i} \right) e^{2\pi it/\tau} \right] \tag{9.27}$$

The sinusoidal head variation is largest at the well and decreases with distance from the well. Discharge variations with a short period damp out at a shorter distance from the well than variations with a longer period. It is noted that taking the real part of the complex function in Eq. (9.25) results in the solution for the case that the sine function in Eq. (9.23) is replaced by a cosine function.

In the example below, the solution is plotted for a sinusoidal well discharge with a period of 1 year in an unconfined aquifer where the transmissivity is again approximated as constant. The function kv of the scipy.special package is used to compute the modified Bessel functions of complex argument, because the functions k0 and k1 only accept real arguments. In the left graph of Figure 9.4, the head variation is plotted at the well and at distances $r = \mu$ and $r = 2\mu$. The amplitude of the head variation at $r = 2\mu$ is negligible compared to the head variation at the well. In the right graph of Figure 9.4, the total flow in the aquifer $Q_{\mathrm{tot}} = 2\pi r Q_r$ is plotted. The amplitude of the total flow equals A at the well and slowly decreases to less than $0.05A$ at $r = 6\mu$.

```
# parameters
T = 100 # transmissivity, m^2/d
S = 1e-1 # storage coefficient, -
tau = 365 # period, d
A = 200 # amplitude of discharge function, m^3/d
rw = 0.3 # radius of well, m
```

```
mu = np.sqrt(T * tau / (S * 2 * np.pi))
print(f'characteristic length mu: {mu:.2f} m')
```

characteristic length mu: 241.02 m

```
# solution
from scipy.special import kv
def hsinwell(r, t):
    hcomplex = -A / (2 * np.pi * T) * kv(0, r * np.sqrt(1j) / mu) * \
               np.exp(2 * np.pi * 1j * t / tau)
    return hcomplex.imag

def Qrsinwell(r, t):
    Qrcomplex = -A / (2 * np.pi) * np.sqrt(1j) / mu * \
                kv(1, r * np.sqrt(1j) / mu) * np.exp(2 * np.pi * 1j * t / tau)
    return Qrcomplex.imag
```

```
# basic plot
t = np.linspace(0, 2 * tau, 100)
plt.subplot(121)
plt.plot(t, hsinwell(rw, t), label='$r=r_w$')
plt.plot(t, hsinwell(mu, t), label='$r=\mu$')
plt.plot(t, hsinwell(2 * mu, t), label='$r=2\mu$')
plt.legend()
plt.subplot(122)
plt.plot(t, Qrsinwell(rw, t) * 2 * np.pi * rw, label='$r=r_w$')
plt.plot(t, Qrsinwell(3 * mu, t) * 2 * np.pi * 3 * mu, 'C3', label='$r=3\mu$')
plt.plot(t, Qrsinwell(6 * mu, t) * 2 * np.pi * 6 * mu, 'C4', label='$r=6\mu$')
plt.legend();
```

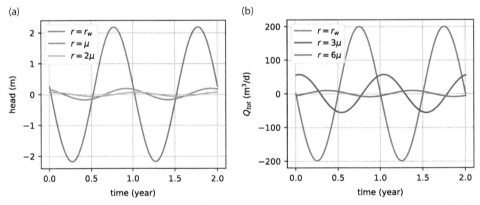

Figure 9.4   Sinusoidal variations vs. time at different distances from a periodic well for the head (left) and Q_{tot} (right).

Exercise 9.7: Use numerical derivatives to verify that Q_r as programmed is indeed $Q_r = -T\partial h/\partial r$ for several values of r and t.

9.3   Wells in a semi-confined aquifer

The solution for steady flow to a well in a semi-confined aquifer was presented in Section 6.4. In this section, the equivalent solution is presented for transient flow. Consider a semi-confined aquifer with transmissivity T and storage coefficient S. The top of the aquifer is formed by a semi-confining layer with resistance c (Figure 6.12). Storage in the semi-confining layer is neglected. The head h^* above the semi-confining layer is fixed to zero. Vertical leakage q_z through the semi-confining layer is computed as Eq. (2.2)

$$q_z = \frac{h - 0}{c} \tag{9.28}$$

so that the governing differential equation for radial flow is

$$\frac{\partial^2 h}{\partial r^2} + \frac{1}{r}\frac{\partial h}{\partial r} = \frac{S}{T}\frac{\partial h}{\partial t} + \frac{h}{cT} \tag{9.29}$$

A well starts pumping with discharge Q at time $t = t_0$. Initially, the head change caused by the well in the aquifer is equal to zero everywhere. The initial and boundary conditions are the same as those for the Theis solutions (9.6, 9.7, and 9.8).

The solution to this problem was first obtained by Hantush and Jacob (1955) and is referred to as the Hantush function

$$h = -\frac{Q}{4\pi T}\int_u^\infty \frac{1}{\tau}\exp\left(-\tau - \frac{r^2}{4\lambda^2\tau}\right)d\tau \tag{9.30}$$

where $\lambda = \sqrt{Tc}$ is the leakage factor and u is defined as for the Theis well

$$u = \frac{Sr^2}{4T(t - t_0)} \tag{9.31}$$

The solution for the case that the head above the semi-confining layer is equal to h^* rather than zero is obtained by simply adding a constant h^* to the solution. Note that this is allowed because $h = h^*$ is a particular solution to the differential equation

$$\frac{\partial^2 h}{\partial r^2} + \frac{1}{r}\frac{\partial h}{\partial r} = \frac{S}{T}\frac{\partial h}{\partial t} + \frac{h - h^*}{cT} \tag{9.32}$$

The seemingly easy integral in the equation for the head (Eq. 9.30) cannot be integrated analytically, nor is it included in the special functions of the scipy.special package; early literature included tabulated values. The easiest way to compute the integral with Python is to use numerical integration using, e.g., the quad function of the scipy.integrate package. The quad function returns a number of items, of which the first one is the computed integral. In the example code below, the head is first plotted as a function of r and then plotted as a function of t (Figure 9.5). The solution for a well in a semi-confined aquifer (Hantush) is compared to the solution for a well in a confined aquifer (Theis). The Hantush and Theis heads are similar for early times when most of the water pumped by the well comes from storage. The Hantush and Theis heads deviate when a significant amount of water pumped by the well comes from leakage through the semi-confining layer. The Hantush solution approaches the steady state solution for a well in a semi-confined aquifer in approximately one day for the parameters used in the example below. The Theis solution does not reach steady state, as explained in Section 9.1.

```
# parameters
T = 200 # transmissivity of aquifer, m^2/d
S = 0.0005 # storage coefficient of aquifer, -
c = 1000 # resistance of leaky layer, d
Q = 800 # discharge of well, m^3/d
```

```
rw = 0.3 # radius of well, m
lab = np.sqrt(c * T) # leakage factor, m
print(f'leakage factor: {lab:.2f} m')
```

leakage factor: 447.21 m

```
# solution
from scipy.integrate import quad

def integrand(tau, r, T, lab):
    return 1 / tau * np.exp(-tau - r ** 2 / (4 * lab ** 2 * tau))

def hantush(r, t, T, S, c, Q):
    lab = np.sqrt(T * c)
    u = S * r ** 2 / (4 * T * t)
    F = quad(integrand, u, np.inf, args=(r, T, lab))[0]
    return -Q / (4 * np.pi * T) * F

hantushvec = np.vectorize(hantush) # vectorized version of hantush function
```

```
# basic plot for one value of r and one value of t
plt.subplot(121)
r = np.linspace(rw, 2 * lab, 100)
t = 6 / 24
plt.plot(r, hantushvec(r, t, T, S, c, Q), label='Hantush')
plt.plot(r, theis(r, t, T, S, Q), label='Theis') # theis func from Section 9.1
plt.legend()
plt.subplot(122)
t = np.linspace(0.01, 2, 100)
r = lab # evaluate head at r=lambda
plt.plot(t, hantushvec(r, t, T, S, c, Q), label='Hantush')
plt.plot(t, theis(r, t, T, S, Q), '--', label='Theis')
plt.legend();
```

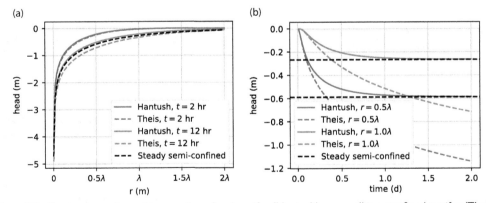

Figure 9.5   Comparison of a well in a semi-confined aquifer (Hantush) vs. a well in a confined aquifer (Theis); head vs. r (left) and head vs. time (right).

Exercise 9.8: Consider a 7-hour pumping test in a leaky aquifer with discharge $Q = 24466$ m^3/d (USBR, 1995, Texas Hill project, Table 9.5). The data in an observation well 24.38 m from the pumping well are given below. First, estimate the transmissivity and storage coefficient approximating the aquifer as confined. Next, estimate the transmissivity, storage coefficient, and resistance of the leaky layer approximating the aquifer as semi-confined. Compute the root-mean squared error, and make a graph of the fit for both models.

```
# data for Exercise 9.8
t = np.array([2, 4, 6, 8, 10, 15, 20, 25, 30, 40, 50, 60, 70, 80, 90, 100,
              110, 120, 150, 180, 210, 240, 270, 300, 360, 420]) # minutes
h = np.array([-0.94, -1.23, -1.54, -1.61, -1.82, -2.05, -2.18, -2.32, -2.43,
              -2.55, -2.63, -2.72, -2.8 , -2.84, -2.89, -2.91, -2.94, -2.97,
              -3.03, -3.07, -3.11, -3.13, -3.15, -3.17, -3.15, -3.18]) # m
```

As an example, the solution for transient flow to a well that starts pumping at $t_0 = 0$ in a semi-confined aquifer is repeated here using a Laplace transform solution following the procedure explained in Section 5.4. The same procedure is used to solve the transient flow problems in the next two sections. The Laplace transform of the differential equation (9.29) is

$$\nabla^2 \overline{h} = \frac{Sp}{T}\overline{h} + \frac{\overline{h}}{cT} = w\overline{h} \qquad (9.33)$$

where $\overline{h}$ is the Laplace-transformed head, p is the Laplace transform parameter, and

$$w = \frac{Scp + 1}{cT} \qquad (9.34)$$

Boundary conditions (9.6) and (9.8) transform to

$$\lim_{r \to 0} -2\pi r \overline{Q}_r = \frac{Q}{p} \qquad (9.35)$$

$$\overline{h}|_{r=\infty} = 0 \qquad (9.36)$$

The radial flow problem in the Laplace domain is mathematically analogous to a steady well in a semi-confined aquifer (Section 6.4), and the solution is

$$\overline{h} = -\frac{Q}{2\pi Tp}K_0\left(r\sqrt{w}\right) \qquad (9.37)$$

Back-transformation to the time domain is conducted numerically using the Stehfest algorithm (see Section 5.4). The example below shows that the Hantush solution using numerical integration of Eq. (9.30) is virtually the same as the solution obtained with the Laplace transform technique and the Stehfest algorithm (Figure 9.6).

```
# Laplace transform solution
from scipy.special import k0
def hbar(p, r, T, S, c, Q):
    w = (S * c * p + 1) / (c * T)
    return -Q / (2 * np.pi * T * p) * k0(r * np.sqrt(w))

t = np.linspace(1e-3, 1, 100)
h =
stehfest(250, t, hbar, T=T, S=S, c=c, Q=Q) # stehfest func from Section 5.4
h_hantush = hantushvec(250, t, T, S, c, Q)
```

```
# basic plot
plt.plot(t, h[0], label='Stehfest')
plt.plot(t, h_hantush, '--', label='Hantush')
plt.legend();
```

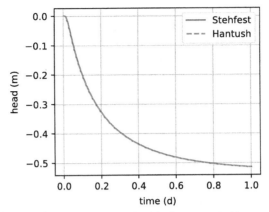

Figure 9.6   Comparison of Laplace transform solution with the numerical integration of the integral in the Hantush solution for a well in a semi-confined aquifer.

9.4   Wells with wellbore storage and skin effect

The previous solutions for transient flow to a well were based on two approximations: The boundary condition at the well was approximated by taking the limit for the well radius going to zero (Eq. 9.6), and the water level in the well is approximated by the head at the well radius in the aquifer. The first approximation neglects the storage of the water inside the well, which is referred to as wellbore storage. The second approximation neglects the resistance of the well screen, which is referred to as the skin effect. Both are taken into account in this section.

Consider radial flow to a well in a confined aquifer with transmissivity T and storage coefficient S. The well starts pumping with discharge Q at time $t = 0$. The radius of the well screen is r_w (see Figure 9.7). The resistance to flow through the well screen is equal to c_w so that the radial component of the discharge vector right at the well screen may be written as

$$Q_r|_{r=r_w} = H\frac{h_w - h|_{r=r_w}}{c_w} \tag{9.38}$$

where h_w is the water level inside the well. Conversely, the water level in the well may be written as

$$h_w = h|_{r=r_w} + \frac{c_w}{H}Q_r|_{r=r_w} \tag{9.39}$$

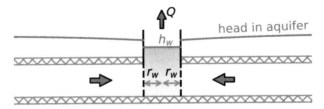

Figure 9.7   Vertical cross section through a well with wellbore storage and skin effect. The head in the aquifer at the well screen is higher than the water level h_w in the well because of the skin effect.

The water balance for water inside the well is

$$A_w\frac{dh_w}{dt} = Q_{in} - Q_{out} \tag{9.40}$$

where A_w is the cross-sectional area of the well screen at the level of the water table in the well ($A_w = \pi r_w^2$ for wells with a constant well radius). The inflow Q_{in} consists of the inflow integrated along the well screen

$$Q_{in} = -2\pi r_w Q_r|_{r=r_w} \tag{9.41}$$

while the outflow is equal to the discharge of the well

$$Q_{out} = Q \qquad t \geq 0 \tag{9.42}$$

Substitution of Eq. (9.39) for h_w, Eq. (9.41) for Q_{in}, and Eq. (9.42) for Q_{out} in the water balance inside the well gives

$$A_w \left[\left.\frac{\partial h}{\partial t}\right|_{r=r_w} + \frac{c_w}{H} \left.\frac{\partial Q_r}{\partial t}\right|_{r=r_w} \right] = -2\pi r_w Q_r|_{r=r_w} - Q \qquad t \geq 0 \tag{9.43}$$

The governing differential equation for flow in the aquifer is the diffusion equation (9.5). The Laplace transform of the diffusion equation is the modified Helmholtz equation

$$\frac{d^2\bar{h}}{dr^2} + \frac{1}{r}\frac{d\bar{h}}{dr} = \frac{Sp}{T}\bar{h} = w\bar{h} \tag{9.44}$$

where

$$w = Sp/T \tag{9.45}$$

The Laplace transform of the water balance inside the well (Eq. 9.43) is

$$A_w p\bar{h}|_{r=r_w} + \left(\frac{A_w c_w p}{H} + 2\pi r_w\right)\overline{Q}_r|_{r=r_w} = -\frac{Q}{p} \tag{9.46}$$

The general solution to the Laplace-transformed diffusion equation in radial coordinates (Eq. 9.44), for the case that $\bar{h}$ equals zero far away, is

$$\bar{h} = B K_0(r\sqrt{w}) \tag{9.47}$$

where B is a parameter to be determined from the boundary conditions. The corresponding discharge vector in the Laplace domain is

$$\overline{Q}_r = -T\frac{d\bar{h}}{dr} = -BT\sqrt{w}K_1(r\sqrt{w}) \tag{9.48}$$

Substitution of the equations for $\bar{h}$ and $\overline{Q}_r$ in the Laplace-transformed water balance (Eq. 9.46) gives

$$B = -\frac{Q}{p}\frac{1}{A_w p K_0(r_w\sqrt{w}) + \left(\frac{A_w c_w p}{H} + 2\pi r_w\right)T\sqrt{w}K_1(r_w\sqrt{w})} \tag{9.49}$$

Back-transformation of the Laplace-transformed solution to the time domain is again carried out using the Stehfest algorithm (see Section 5.4). Note that for the case that wellbore storage is neglected, the cross-sectional area A_w is set to zero in the water balance equation (9.40), which means that ($\pi r_w^2 = 0$) and B reduces to

$$B = -\frac{Q}{p}\frac{1}{2\pi r_w T\sqrt{w}K_1(r_w\sqrt{w})} \tag{9.50}$$

which is the solution when the boundary condition at the well screen is

$$-2\pi r_w Q_r|_{r=r_w} = Q \qquad t \geq 0 \tag{9.51}$$

In the example below, the head change is compared for a well without wellbore storage (boundary condition 9.51), a well with wellbore storage but no skin effect, and a well with both wellbore storage and skin effect (Figure 9.8). For the values of the example, wellbore storage and skin effect have an effect on the head during the first few hours of pumping ($t < 0.1$ d). For the case with skin effect, the head inside the well starts to differ from the head just outside the well after a few minutes of pumping ($t > 0.002$ d).

```
# parameters
T = 200 # transmissivity, m^2/d
S = 0.005 # storage coefficient of aquifer, -
H = 10 # aquifer thickness, m
Q = 800 # discharge of well, m^3/d
rw = 0.3 # well radius, m
cw = 0.02 # resistance of well screen, d
```

```
# solution
from scipy.special import k0, k1
def hbar_wbs_res(p, r, T, S, Q, rw, cw, H): # hbar
    sqrtw = np.sqrt(S * p / T)
    Aw = np.pi * rw ** 2
    B = -Q / p * 1 / (p * Aw * k0(rw * sqrtw) +
                     (2 * np.pi * rw + Aw * p * cw / H) *
                     T * sqrtw * k1(rw * sqrtw))
    hbar = B * k0(r * sqrtw)
    return hbar

def Qrbar_wbs_res(p, r, T, S, Q, rw, cw, H): # Qrbar
    sqrtw = np.sqrt(S * p / T)
    Aw = np.pi * rw ** 2
    B = -Q / p * 1 / (p * Aw * k0(rw * sqrtw) +
                     (2 * np.pi * rw + Aw * p * cw / H) *
                     T * sqrtw * k1(rw * sqrtw))
    Qrbar = T * sqrtw * B * k1(r * sqrtw)
    return Qrbar

def hwbar_wbs_res(p, r, T, S, Q, rw, cw, H): # hbar inside well
    # r is not used, but Stehfest routine requires it
    hrw = hbar_wbs_res(p, rw, T, S, Q, rw, cw, H)
    Qrw = Qrbar_wbs_res(p, rw, T, S, Q, rw, cw, H)
    return cw * Qrw / H + hrw

# import stehfest_coef and stehfest from Section 5.4
t = np.logspace(-3, 0, 100)
ht = htheis(rw, t, T, S, Q) # function from Section 9.1
hw = stehfest(0, t, hwbar_wbs_res, T=T, S=S, Q=Q, rw=rw, cw=cw, H=H)
hwbs = stehfest(rw, t, hbar_wbs_res, T=T, S=S, Q=Q, rw=rw, cw=0, H=H)
hwbsres = stehfest(rw, t, hbar_wbs_res, T=T, S=S, Q=Q, rw=rw, cw=cw, H=H)
```

```
# plot of head at well
plt.semilogx(t, ht, label='x wbs, x skin')
plt.semilogx(t, hwbs[0], label=f'v wbs, x sin')
```

```
plt.semilogx(t, hwbsres[0], label=f'v wbs, v skin')
plt.semilogx(t, hw[0], '--', label=f'in well, v wbs, v skin')
plt.legend();
```

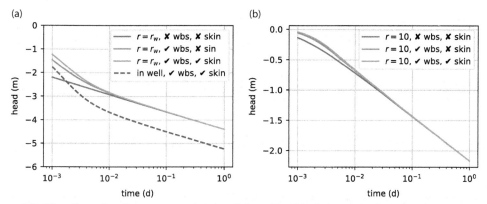

Figure 9.8 The effect of wellbore storage (wbs) and skin effect (skin) during the first day of pumping; head at the well (left) and head at $r = 10$ m (right).

Exercise 9.9: Analyze the pumping test at Oude Korendijk (Section 9.1) taking into account wellbore storage using a well radius of 0.3 m, but no skin effect. Compare your answer to the results of Section 9.1.

9.5   Wells in a two-aquifer system

The solution for steady flow to a well in the bottom aquifer of a two-aquifer system was presented in Section 6.5. In this section, the transient equivalent is presented. The solution is used to investigate the transient head response in a two-aquifer system. When the top aquifer is unconfined, the head response in the top aquifer will be significantly slower than the head response in the bottom aquifer, also referred to as the delayed response of the water table. The solution is quite complicated as it involves solving a system of two coupled differential equations using the Laplace transform method. Although the equations are a bit longer, the Python script of the solution is relatively short.

Consider transient flow in a two-aquifer system. The top aquifer has index 0, and the bottom aquifer has index 1. The transmissivity of each aquifer is again approximated as constant in both space and time. The storage coefficients of the top and bottom aquifers are S_0 and S_1, respectively. The two aquifers are separated by a leaky layer with resistance c (Figure 6.14). The storage in the leaky layer is neglected. The transient equivalent of the system of two linked differential equations (6.63) is

$$
\begin{aligned}
\nabla^2 h_0 &= \frac{S_0}{T_0}\frac{\partial h_0}{\partial t} + \frac{h_0 - h_1}{cT_0} \\
\nabla^2 h_1 &= \frac{S_1}{T_1}\frac{\partial h_1}{\partial t} + \frac{h_1 - h_0}{cT_1}
\end{aligned}
\tag{9.52}
$$

A well starts pumping with a discharge Q at time $t = 0$ in aquifer 1 (the bottom aquifer). The boundary conditions are

$$
\begin{aligned}
&\lim_{r \to 0} -2\pi r Q_{r_0} = 0 \\
&\lim_{r \to 0} -2\pi r Q_{r_1} = Q \qquad t \geq 0
\end{aligned}
\tag{9.53}
$$

$$
h_0|_{r=\infty, t} = h_1|_{r=\infty, t} = 0
\tag{9.54}
$$

and the initial condition is

$$h_0|_{r,t=0} = h_1|_{r,t=0} = 0 \tag{9.55}$$

A solution is obtained using Laplace transforms, as explained in Section 5.4. The solution procedure is a bit more complicated, however, as the system of two linked differential equations needs to be broken down into the solution of two unlinked differential equations. This unlinking is more difficult than for steady flow (Section 6.4) and requires the solution of an eigenvalue problem.

The Laplace-transformed system of differential equations is

$$\nabla^2 \overline{h}_0 = \frac{S_0 p}{T_0} \overline{h}_0 + \frac{\overline{h}_0 - \overline{h}_1}{cT_0}$$
$$\nabla^2 \overline{h}_1 = \frac{S_1 p}{T_1} \overline{h}_1 + \frac{\overline{h}_1 - \overline{h}_0}{cT_1} \tag{9.56}$$

which may be written in matrix form as

$$\nabla^2 \begin{pmatrix} \overline{h}_0 \\ \overline{h}_1 \end{pmatrix} = \begin{pmatrix} \frac{S_0 cp+1}{cT_0} & \frac{-1}{cT_0} \\ \frac{-1}{cT_1} & \frac{S_1 cp+1}{cT_1} \end{pmatrix} \begin{pmatrix} \overline{h}_0 \\ \overline{h}_1 \end{pmatrix} \tag{9.57}$$

The matrix on the right-hand side of the matrix equation is called the system matrix $\mathbf{A}$ and is written as

$$\mathbf{A} = \begin{pmatrix} \frac{S_0 cp+1}{cT_0} & \frac{-1}{cT_0} \\ \frac{-1}{cT_1} & \frac{S_1 cp+1}{cT_1} \end{pmatrix} = \mathbf{VWV}^{-1} \tag{9.58}$$

where $\mathbf{W}$ is a diagonal matrix with the eigenvalues of matrix $\mathbf{A}$ on the diagonal

$$\mathbf{W} = \begin{pmatrix} w_0 & 0 \\ 0 & w_1 \end{pmatrix} \tag{9.59}$$

and $\mathbf{V}$ is a matrix with the corresponding eigenvectors $\vec{u} = \begin{pmatrix} u_0 \\ u_1 \end{pmatrix}$ and $\vec{v} = \begin{pmatrix} v_0 \\ v_1 \end{pmatrix}$ as its columns

$$\mathbf{V} = \begin{pmatrix} u_0 & v_0 \\ u_1 & v_1 \end{pmatrix} \tag{9.60}$$

The boundary conditions in the Laplace domain are

$$\lim_{r \to 0} -2\pi r \overline{Q}_{r_0} = 0$$
$$\lim_{r \to 0} -2\pi r \overline{Q}_{r_1} = \frac{Q}{p} \tag{9.61}$$

$$\overline{h}_0|_{r=\infty} = \overline{h}_1|_{r=\infty} = 0 \tag{9.62}$$

The solution in the Laplace domain is

$$\begin{pmatrix} \overline{h}_0 \\ \overline{h}_1 \end{pmatrix} = -\frac{Q}{2\pi T_1 p} \left[b_0 \, \mathrm{K}_0 \left(r\sqrt{w_0} \right) \begin{pmatrix} u_0 \\ u_1 \end{pmatrix} + b_1 \, \mathrm{K}_0 \left(r\sqrt{w_1} \right) \begin{pmatrix} v_0 \\ v_1 \end{pmatrix} \right] \tag{9.63}$$

where b_0 and b_1 are the solution to the following system of two linear equations

$$\begin{pmatrix} u_0 & v_0 \\ u_1 & v_1 \end{pmatrix} \begin{pmatrix} b_0 \\ b_1 \end{pmatrix} = \begin{pmatrix} 0 \\ 1 \end{pmatrix} \tag{9.64}$$

The Python implementation of the solution makes use of the `eig` function of the `lingalg` subpackage of numpy. The `eig` function returns the array $[w_0, w_1]$ of eigenvalues and the two-dimensional array **V** with the eigenvectors.

In the example below, the top aquifer is unconfined so that the storage coefficient S_0 represents the phreatic storage of the top aquifer. The head is computed and plotted in an observation well near the well (Figure 9.9). The head responds very quickly in the bottom aquifer, but it takes approximately one day before the head in the unconfined top aquifer starts to respond. The head in the bottom aquifer starts to go down again when the head in the top aquifer starts to go down. This delayed response of the water table can be seen especially well in a semi-log plot. For early times (in this case, for the first day), the head response is similar to a Hantush well, as the head in the top aquifer varies very little at early time.

```
# parameters
T0 = 100 # transmissivity top aquifer, m^2/d
S0 = 0.1 # phreatic storage top aquifer, -
T1 = 200 # transmissivity bottom aquifer, m^2/d
S1 = 0.001 # storage coefficient bottom aquifer, -
c = 500 # resistance of leaky layer, d
Q = 800 # discharge of well, m^3/d
rw = 0.3 # radius of well, m
```

```
# solution
from scipy.special import k0
def hmaqbar(plist, r, layer):
    hbar = np.zeros(len(plist))
    for i, p in enumerate(plist):
        A = np.array([[(p * c * S0 + 1) / (c * T0), -1 / (c * T0)],
                      [-1 / (c * T1), (p * c * S1 + 1) / (c * T1)]])
        w, v = np.linalg.eig(A)
        b = np.linalg.solve(v, [0, 1])
        hbar[i] = -Q / (2 * np.pi * T1 * p) * (
            b[0] * v[layer, 0] * k0(r * np.sqrt(w[0])) +
            b[1] * v[layer, 1] * k0(r * np.sqrt(w[1])))
    return hbar

r = 250
t = np.logspace(-3, 2, 100)
h0 = stehfest(r, t, hmaqbar, layer=0) # stehfest function from Section 5.4
h1 = stehfest(r, t, hmaqbar, layer=1)
h1_hantush = hantushvec(r, t, T1, S1, c, Q) # hantushvec from Section 9.3
```

```
# basic plot semi-log
plt.semilogx(t, h0[0])
plt.semilogx(t, h1[0])
plt.semilogx(t, h1_hantush);
```

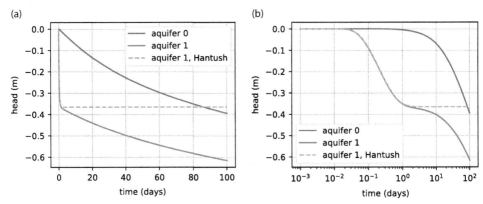

Figure 9.9 Drawdown in an observation well near a pumping well in the bottom aquifer of a two-aquifer
system. The top aquifer is unconfined; linear time axis (left) and log-scaled time axis (right).

Chapter 10

Steady two-dimensional flow in the vertical plane

So far in this book, solutions were derived based on the Dupuit approximation. This means that the resistance to flow in the vertical direction is neglected in an aquifer. As a result, the head does not vary in the vertical direction. The vertical component of flow inside an aquifer is computed from continuity of flow. This is a very good approximation for many groundwater problems where the flow is predominantly horizontal. But there are flow situations, of course, where vertical flow in an aquifer does play an important role. Such problems may include flow to partially penetrating streams or wells, flow in aquifers with an irregular top or base, and flow toward the bottom of lakes or oceans. For such problems, the question arises whether the Dupuit approximation is still a good approximation, especially when the vertical hydraulic conductivity is (much) smaller than the horizontal hydraulic conductivity. In this chapter, solutions are presented for two-dimensional flow in a vertical cross section where the Dupuit approximation is not adopted. The range of applicability of Dupuit solutions is assessed by comparing several two-dimensional solutions to equivalent Dupuit solutions.

Solutions for steady two-dimensional flow in the vertical plane tend to be derived with advanced mathematical techniques such as conformal mapping, the hodograph method, or separation of variables. The solutions presented in this chapter are all written in terms of a complex specific discharge potential, which is introduced in the following. All solutions are presented rather than derived.

Two-dimensional flow in the vertical plane is formulated in terms of the specific discharge vector with components q_x in the horizontal direction and q_z in the vertical direction. Continuity for steady flow in the vertical x, z plane is (see Eq. 1.28)

$$\frac{\partial q_x}{\partial x} + \frac{\partial q_z}{\partial z} = 0 \tag{10.1}$$

where the z-axis points vertically upward. The components of the specific discharge vector for an aquifer with isotropic and homogeneous hydraulic conductivity k may be written as

$$q_x = -k\frac{\partial h}{\partial x} = -\frac{\partial \phi}{\partial x} \qquad q_z = -k\frac{\partial h}{\partial z} = -\frac{\partial \phi}{\partial z} \tag{10.2}$$

where

$$\phi = kh \tag{10.3}$$

is the specific discharge potential. The dimensions of the specific discharge potential ϕ are L^2/T, which is in contrast to the discharge potential Φ used in previous chapters, which has dimensions L^3/T. The negative gradient of the specific discharge potential ϕ results in the specific discharge vector $\vec{q}$, while the negative gradient of the discharge potential Φ results

DOI: 10.1201/9781315206134-10

in the discharge vector $\vec{Q}$. Substitution of the components of the specific discharge vector (Eq. 10.2) into the continuity equation (Eq. 10.1) gives Laplace's equation

$$\frac{\partial^2 \phi}{\partial x^2} + \frac{\partial^2 \phi}{\partial z^2} = 0 \tag{10.4}$$

For two-dimensional steady flow in the vertical plane governed by Laplace's equation, there is also a stream function, which is written as ψ [L^2/T], as was done for flow in a vertical cross section in Chapters 1–4. The specific discharge potential and stream function are combined in a complex potential

$$\omega = \phi + i\psi \tag{10.5}$$

The complex specific discharge may be obtained from ω as

$$q_x - iq_z = -\frac{d\omega}{d\zeta} \tag{10.6}$$

where $\zeta = x + iz$ is the complex coordinate in a vertical cross section.

10.1   Vertical anisotropy

The hydraulic conductivity is often anisotropic, where the vertical component of the hydraulic conductivity is smaller than the horizontal component. One of the main reasons for vertical anisotropy is that aquifers often consist of alternating layers of somewhat coarser and somewhat finer material. The average value of the hydraulic conductivity in the horizontal direction is dictated by the hydraulic conductivity of the coarser layers, while the average value in the vertical direction is dictated by the hydraulic conductivity of the finer layers. In practice, it is popular to use an anisotropy factor of 10 (the vertical hydraulic conductivity is 10 times smaller than the horizontal hydraulic conductivity), but a good reason for this number does not exist. For the case that the hydraulic conductivity is anisotropic with k_x in the x-direction and k_z in the z-direction, Darcy's law becomes

$$q_x = -k_x \frac{\partial h}{\partial x} \qquad q_z = -k_z \frac{\partial h}{\partial z} \tag{10.7}$$

Substitution of these equations for q_x and q_z in the continuity equation (10.1) gives

$$\frac{\partial}{\partial x} \left(k_x \frac{\partial h}{\partial x} \right) + \frac{\partial}{\partial z} \left(k_z \frac{\partial h}{\partial z} \right) = 0 \tag{10.8}$$

This equation may be transformed into Laplace's equation by a change of variables

$$\tilde{x} = x\sqrt{k_z/k_x} \qquad \tilde{z} = z \tag{10.9}$$

which gives

$$\frac{\partial^2 h}{\partial \tilde{x}^2} + \frac{\partial^2 h}{\partial \tilde{z}^2} = 0 \tag{10.10}$$

Hence, in the $\tilde{x}, \tilde{z}$ domain, the head is governed by Laplace's equation. The specific discharge in the transformed domain can be evaluated with Darcy's law using the isotropic hydraulic conductivity $\tilde{k}$ defined as

$$\tilde{k} = \sqrt{k_x k_z} \tag{10.11}$$

The use of $\tilde{k}$ ensures that the net discharge in the isotropic transformed domain is the same as the net discharge in the original anisotropic domain. For example, the net flux Q_{net} across a small section Δz in the anisotropic domain equals

$$Q_{net} = -k_x \frac{\partial h}{\partial x} \Delta z \qquad (10.12)$$

and is equal to the net flux across the same section in the transformed isotropic domain

$$\tilde{Q}_{net} = -\tilde{k}\frac{\partial h}{\partial \tilde{x}}\Delta \tilde{z} = -\sqrt{k_x k_z}\frac{\partial h}{\partial x}\frac{1}{\sqrt{k_z/k_x}}\Delta z = -k_x\frac{\partial h}{\partial x}\Delta z = Q_{net} \qquad (10.13)$$

The same holds in the z-direction.

In summary, an anisotropic problem may be solved with standard methods for Laplace's equation, provided that the x-direction of the flow field is squeezed according to Eq. (10.9) and the isotropic hydraulic conductivity (Eq. 10.11) is used in the transformed domain. A complex specific discharge potential in the transformed potential may be defined as $\tilde{\omega} = \tilde{\phi} + i\tilde{\psi}$, where $\tilde{\phi} = \tilde{k}h$ and $\tilde{\psi} = \psi$.

As a simple example, consider a rectangular domain with length L and height B (Figure 10.1). The head at the origin in the lower left-hand corner equals zero. The flow is uniform, and the head gradients in the x- and z-directions are equal: $\partial h/\partial x = \partial h/\partial z = a$. As such, the head varies linearly along the boundaries (Figure 10.1). The hydraulic conductivity in the x- and z-directions is k_x and k_z, respectively.

The flow is uniform in both the anisotropic physical domain and the isotropic transformed domain. In the physical domain, $q_x = -k_x a$ and $q_z = -k_z a$. In the transformed domain, $\tilde{q}_x$ and $\tilde{q}_z$ are

$$\tilde{q}_x = -\tilde{k}\frac{\partial h}{\partial \tilde{x}} = -\tilde{k}\frac{\partial h}{\partial x}\frac{\partial x}{\partial \tilde{x}} = -\frac{\tilde{k}a}{\sqrt{k_z/k_x}} \qquad (10.14)$$

$$\tilde{q}_z = -\tilde{k}a \qquad (10.15)$$

The solution for the (uniform) complex specific discharge potential in the transformed domain is Eq. (7.19)

$$\tilde{\omega} = -(\tilde{q}_x - \tilde{q}_z i)\tilde{\zeta} \qquad (10.16)$$

where $\tilde{\zeta} = \tilde{x} + i\tilde{z}$ is the complex coordinate in the transformed domain.

An example flow field is shown in Figure 10.1. The flow field is shown in both the transformed domain and the problem domain. Note that the streamlines and head contours intersect at right angles in the transformed domain, where the hydraulic conductivity $\tilde{k}$ is isotropic, but don't intersect at right angles in the problem domain, where the hydraulic conductivity is anisotropic.

```
# parameters
L = 1000 # length of domain, m
B = 300 # width of domain, m
kx = 5 # hydraulic conductivity in x-direction, m/d
kz = 1 # hydraulic conductivity in z-direction, m/d
a = -0.002 # head gradient in x- and z-directions
```

```
# solution
kt = np.sqrt(kx * kz)
qxt = -kt * a / np.sqrt(kz / kx)
```

```
qzt = -kt * a
xg, zg = np.meshgrid(np.linspace(0, L, 10), np.linspace(0, B, 10))
xt = xg * np.sqrt(kz / kx)
zt = zg
zetat = xt + 1j * zt
omt = -(qxt - 1j * qzt) * zetat
h = omt.real / kt
psi = omt.imag
```

```
# basic flow net
plt.subplot(121, aspect=1) # problem domain
plt.contour(xg, zg, h, 10)
plt.contour(xg, zg, psi, 10);
plt.subplot(122, aspect=1) # transformed domain
plt.contour(xt, zt, h, 20)
plt.contour(xt, zt, psi, 20);
```

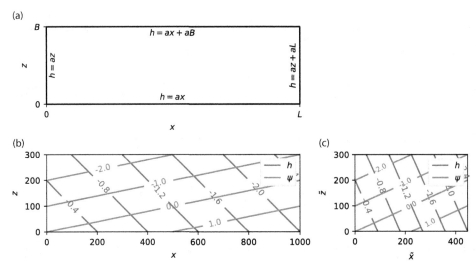

Figure 10.1   Example for uniform flow in an anisotropic aquifer; boundary conditions (upper left), flow net in squeezed transformed domain (lower right), and flow net in problem domain (lower left).

10.2   Flow to a partially penetrating stream

A typical approximation in the development of analytical solutions is that a stream fully cuts through the aquifer. Most streams, however, penetrate only part of the aquifer. Such streams are called partially penetrating streams. In this section, the flow field around a narrow partially penetrating stream is investigated. A solution is presented for two-dimensional flow in the vertical plane and compared to the Dupuit solution.

The flow system consists of a confined aquifer with hydraulic conductivity k and thickness H and is penetrated by a narrow stream over a distance d (Figure 10.2). The water level in the stream equals h_s. The width of the stream is approximated as negligible (even though it has a small width in the figure). The flow near infinity on the left equals $Q_x = U_L$, while the flow near infinity on the right equals $Q_x = U_R$, so that the boundary conditions are

$$\lim_{x \to -\infty} Q_x = U_L \tag{10.17}$$

$$\lim_{x \to \infty} Q_x = U_R \tag{10.18}$$

$$h|_{x=0,-d \le z \le 0} = h_s \tag{10.19}$$

The solution to the stated problem is obtained with conformal mapping. A conformal mapping is a coordinate transformation (a change of variables) for which the Laplace equation is invariant. Some boundary conditions, including a specified head boundary and an impermeable boundary, are also invariant to a conformal transformation. The utility of a conformal mapping is to simplify the problem domain without altering the differential equation or boundary conditions. An analytic solution is then obtained in the simplified domain. The problem domain is also referred to as the physical plane, and the simplified transformed domain is called the reference plane.

A conformal mapping solution is often written in parametric form as $\omega(\tau)$ and $\zeta(\tau)$, where τ is the reference plane. Here, the reference plane $\tau = \xi + i\eta$ is chosen as the quarter plane $\xi \geq 0$ and $\eta \geq 0$, as shown in Figure 10.2 (note that ξ (xi) and ζ (zeta) look similar, but ξ has a double squiggle and ζ just a single squiggle). The top and bottom of the aquifer correspond to the positive real axis of τ (note the corresponding green, red, and purple sections in the ζ and τ planes), while the partially penetrating stream corresponds to the positive imaginary axis in the τ plane (the blue section).

One of the characteristics of a conformal mapping is that there are three degrees of freedom to make a mapping unique; the locations of three boundary points can be chosen in the reference plane. Here, it is chosen that point 1 corresponds to $\tau = 1$, point 2 corresponds to $\tau = \infty$, and point 4 corresponds to $\tau = 0$. Point 3 is located on the imaginary axis in the τ plane, and point 5 is located between $\tau = 0$ and $\tau = 1$ on the real axis in the τ plane, but their exact locations are determined by the mapping.

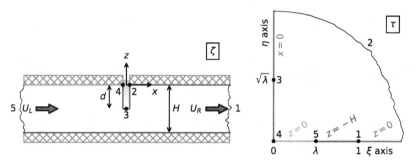

Figure 10.2 Flow to a partially penetrating stream; physical ζ plane (left) and the reference τ plane (right) with the boundary conditions for $\zeta = x + iz$.

The solution for ζ as a function of τ, referred to in conformal mapping jargon as the map of τ onto ζ, is

$$\zeta = x + iz = -\frac{H}{\pi} \ln \left(\frac{\tau - 1}{\tau + 1} \right) + \frac{H}{\pi} \ln \left(\frac{\tau - \lambda}{\tau + \lambda} \right) \tag{10.20}$$

where

$$\lambda = \left(-\mu + \sqrt{\mu^2 + 1} \right)^2 \tag{10.21}$$

and

$$\mu = \tan \left(\frac{\pi d}{2H} \right) \tag{10.22}$$

The parameters μ and λ depend on the geometry of the problem (i.e., the relative depth of penetration of the stream d/H) and not on the flow. For this case, it is possible to invert $\zeta(\tau)$ to obtain $\tau(\zeta)$ as

$$\tau = -\frac{(Z+1)(\lambda-1)}{2(Z-1)} + \frac{1}{2}\sqrt{\left(\frac{(Z+1)(\lambda-1)}{Z-1}\right)^2 + 4\lambda} \qquad (10.23)$$

where

$$Z = e^{\pi\zeta/H} \qquad (10.24)$$

It can be checked analytically that Eq. (10.20) fulfills all the boundary conditions. Note that the solution for ζ consists of two log terms. Each log term is mathematically equivalent to a well and its image across the imaginary axis in the reference plane (compare Section 6.2). As such, the real part of ζ (which is x) is constant along the imaginary axis in the τ plane and the imaginary part (which is z) is piecewise constant along the real axis in the τ plane. It can also be verified numerically that all boundary conditions are met and the solution is implemented correctly in Python by evaluation of ζ at points along the boundary in the τ plane. An example for a coordinate between points 1 and 2 in the τ is shown below. The section between points 1 and 2 corresponds to $z = 0$ and $x \geq 0$ (see Figure 10.2).

```
# parameters
k = 10 # hydraulic conductivity, m/d
H = 10 # thickness of aquifer, m
d = 4 # depth of stream, m
UL = 0.2 # uniform flow far to the left, m^2/d
UR = -0.2 # uniform flow far to the right, m^2/d
hs = 0 # water level in stream
```

```
# solution
def zetafunc(tau, H=H, d=d):
    mu = np.tan(np.pi * d / (2 * H))
    lab = (-mu + np.sqrt(mu**2 + 1))**2
    z = -H / np.pi * np.log((tau - 1) / (tau + 1)) + \
        H / np.pi * np.log((tau - lab) / (tau + lab))
    return z

def taufunc(zeta, H=H, d=d):
    mu = np.tan(np.pi * d / (2 * H))
    lab = (-mu + np.sqrt(mu**2 + 1))**2
    Z = np.exp(np.pi * zeta / H)
    tau = (1 - lab) * (Z + 1) / (2 * (Z - 1)) + 0.5 * np.sqrt(
        ((lab - 1) * (Z + 1) / (Z - 1)) ** 2 + 4 * lab)
    return tau
```

```
# verification
tau = 1.5 + 0j
zeta = zetafunc(tau)
print(f'given tau between points 1 and 2: {tau:.2f}')
print(f'zeta between points 1 and 2: {zeta:.2f}')
print(f'tau as function of zeta: {taufunc(zeta):.2f}')
```

```
given tau between points 1 and 2: 1.50+0.00j
zeta between points 1 and 2: 4.01+0.00j
tau as function of zeta: 1.50+0.00j
```

The solution for the complex potential is obtained next. The stream function is chosen to equal $\psi = 0$ along the top of the aquifer to the right of the stream (the purple section between points 1 and 2) so that $\psi = U_R$ along the bottom of the aquifer (the red section between points 5 and 1) and $\psi = U_R - U_L$ along the top of the aquifer to the left of the stream (the green section between points 5 and 4). The boundary conditions for ω in the reference plane τ are shown in the left graph of Figure 10.3.

The solution for $\omega(\tau)$ in the reference plane may, for this case, be constructed directly in the reference plane with the method of images. Point 5, which is at infinity to the left, in the ζ plane corresponds to the point $\xi = \lambda$ in the τ plane. There is an amount U_L flowing from infinity toward the partially penetrating stream. Hence, the solution for ω needs an injection well with injection rate $2U_L$ at point 5. The factor of 2 is needed as only half the water that is injected flows into the upper half-plane ($\eta \geq 0$), while the other half flows into the lower half-plane ($\eta \leq 0$). Similarly, point 1, which is at infinity to the right in the ζ plane, corresponds to the point $\xi = 1$ in the τ plane. An amount U_R is taken out at point 1, which means that a well with discharge $2U_R$ is placed at point 1. Image wells are used to make sure that the potential is equal to $\phi_s = kh_s$ along the imaginary axis (corresponding to the stream) so that the function for ω becomes

$$\omega = \phi + i\psi = -\frac{U_L}{\pi} \ln\left(\frac{\tau - \lambda}{\tau + \lambda}\right) + \frac{U_R}{\pi} \ln\left(\frac{\tau - 1}{\tau + 1}\right) + \phi_s \qquad (10.25)$$

The potential may be evaluated at a point (ζ) by first computing τ for the specified value of ζ with Eq. (10.23), followed by the evaluation of ω with Eq. (10.25).

An example is presented for the case that the flow is the same from both sides of the stream. A flow net in the τ plane is shown in the right graph of Figure 10.3. A flow net in the problem domain is shown in Figure 10.4. For a fully penetrating stream, the head contours $h = 0.02$ and $h = 0.04$ m are 10 and 20 m from the stream, respectively (these contours are shown in black in Figure 10.4). Note that the vertical head variation is negligible at approximately one aquifer thickness from the partially penetrating stream.

```
# solution
def omegafunc(tau, UL=UL, UR=UR, H=H, d=d, k=k, hs=0):
    mu = np.tan(np.pi * d / (2 * H))
    lab = (-mu + np.sqrt(mu**2 + 1))**2
    om = -UL / np.pi * np.log((tau - lab) / (tau + lab)) + \
          UR / np.pi * np.log((tau - 1) / (tau + 1)) + k * hs
    return om
```

```
# omega transformed domain
xi, eta = np.meshgrid(np.linspace(0, 2, 100), np.linspace(0, 1.5, 100))
om_ref_plane = omegafunc(xi + eta * 1j)
# omega physical domain
xg, zg = np.meshgrid(np.linspace(-4 * H, 4 * H, 400), np.linspace(-H, 0, 100))
taug = taufunc(xg + zg * 1j)
omega = omegafunc(taug)
h = omega.real / k
```

```
# basic plot flow net in transformed domain (right graph)
plt.subplot(111, aspect=1)
plt.contour(xi, eta, om_ref_plane.real, 20, colors='C0')
plt.contour(xi, eta, om_ref_plane.imag, 20, colors='C1')
```

```
# basic flow net
plt.subplot(111, aspect=1, xlim=(-40, 10))
plt.contour(xg, zg, h, np.arange(0, 0.1, 0.0025), colors='C0')
plt.contour(xg, zg, omega.imag, np.arange(-0.4, 0.4, 0.025), colors='C1');
```

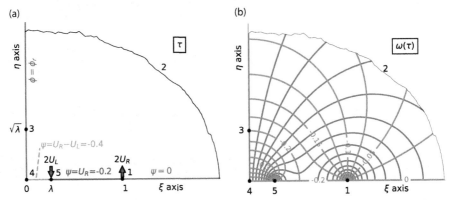

Figure 10.3 Boundary conditions for ω in the reference plane (left) and flow net in the reference plane (right).

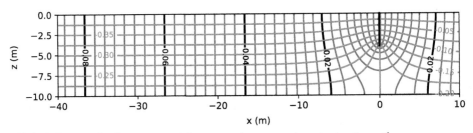

Figure 10.4 A flow net for flow to a partially penetrating stream in an isotropic aquifer.

The solution for an anisotropic hydraulic conductivity is obtained through the application of the procedure outlined in the previous section. The flow net is quite different for the case that the vertical hydraulic conductivity is ten times smaller than the horizontal hydraulic conductivity, as shown in Figure 10.5. The vertical head variation is negligible at approximately $H\sqrt{k_x/k_z} \approx 3H$ from the stream.

```
# anisotropic parameters
kx = 10 # m/d
kz = 1 # m/d
```

```
# anisotropic solution
ktilde = np.sqrt(kx * kz)
xgtilde = xg * np.sqrt(kz / kx)
zgtilde = zg
omega_aniso = omegafunc(taufunc(xgtilde + zgtilde * 1j))
h_aniso = omega_aniso.real / ktilde
psi_aniso = omega_aniso.imag
```

```
# basic flow net
plt.subplot(111, aspect=1, xlim=(-40, 10))
plt.contour(xg, zg, h_aniso, np.arange(0, 0.2, 0.002), colors='C0')
plt.contour(xg, zg, psi_aniso, np.arange(-0.4, 0.4, 0.02), colors='C1');
```

The presented solution is compared to the Dupuit solution for one-dimensional flow toward a stream. First, the Dupuit solution is presented for the case that the head in the aquifer at the stream, h_0, is equal to the head in the stream, h_s, over the entire aquifer thickness

$$h_0 = h_s \tag{10.26}$$

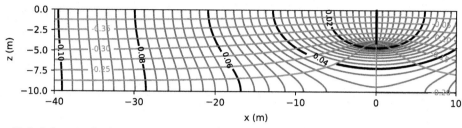

Figure 10.5 A flow net for flow to a partially penetrating stream in an anisotropic aquifer with $k_z = k_x/10$.

The boundary conditions to the left and right are Eqs. (10.17) and (10.18), respectively. The Dupuit solution is

$$h = -U_L x/T + h_0 \quad x \leq 0$$
$$h = -U_R x/T + h_0 \quad x \geq 0 \tag{10.27}$$

The Dupuit solution is compared to the two-dimensional solution for the example with anisotropic hydraulic conductivity in Figure 10.6. Note that away from the stream the difference between the Dupuit solution and the head at the top and bottom of the aquifer becomes constant.

```
# Dupuit solution
L = 40 # distance over which Dupuit solution is computed
h0 = 0
xdupuit = [-L, 0, L]
hdupuit1 = [UL * L / (k * H) + h0, h0, -UR * L / (k * H) + h0]
```

```
# basic plot
plt.plot(xg[0], h_aniso[-1], 'C0', label='$z=0$')
plt.plot(xg[0], h_aniso[0], 'C0--', label='$z=-H$')
plt.plot(xdupuit, hdupuit1, 'C3', label='Dupuit w/o $C$')
plt.legend(loc='lower right');
```

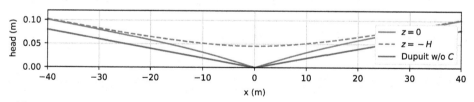

Figure 10.6 Head vs. x at the top and bottom of an anisotropic aquifer with $k_z = k_x/10$ compared to the Dupuit solution with $h|_{x=0} = 0$.

Streamlines converge when water approaches a partially penetrating stream, which results in additional head loss that is not taken into account in the Dupuit solution. The additional head loss can be taken into account in the Dupuit solution by replacing the boundary condition at the stream by

$$U_L - U_R = C(h_0 - h_s) \tag{10.28}$$

where $U_L - U_R$ is the outflow into the stream and C is the effective conductance. The effective conductance C is computed as

$$C = (U_L - U_R)/(\bar{h}_0 - h_s) \tag{10.29}$$

where $\bar{h}_0$ is the average head in the aquifer at $x = 0$ and is obtained from numerical integration of the two-dimensional solution. The effective conductance is a function of the geometry of the problem (H and d) and the hydraulic conductivity (k_x and k_z), but not of the flow toward the stream ($U_R - U_L$). Once C is known, the head h_0 at $x = 0$ in the Dupuit solution can be computed for any combination of U_R and U_L as

$$h_0 = h_s + (U_L - U_R)/C \qquad (10.30)$$

In the code cell below, the effective conductance C is computed from the average head $\bar{h}_0$ for the anisotropic example presented above. The Dupuit solution with the effective conductance is shown in Figure 10.7 and indeed matches the head in the aquifer at a distance of ~$3H$ from the stream.

```
# computation of C
def head0(z, UL=UL, UR=UR, H=H, d=d):
    zeta = 0 + z * 1j
    tau = taufunc(zeta, H=H, d=d)
    return omegafunc(tau, UL=UL, UR=UR, H=H, d=d).real / np.sqrt(kx * kz)

from scipy.integrate import quad
h0avg = quad(head0, a=-H, b=0, args=(UL, UR))[0] / H
C = (UL - UR) / (h0avg - 0)
print(f'Effective conductance: C={C:.2f} m^2/d')
print(f'Dupuit head h0={h0avg:.3f} m for UL={UL} m^2/d and UR={UR} m^2/d')
```

```
Effective conductance: C=18.70 m^2/d
Dupuit head h0=0.021 m for UL=0.2 m^2/d and UR=-0.2 m^2/d
```

```
# Dupuit solution with effective conductance
h0 = hs + (UL - UR) / C
hdupuit2 = [UL * L / (k * H) + h0, h0, -UR * L / (k * H) + h0]
```

```
# basic plot
plt.plot(xg[0], h_aniso[-1], 'C0', label='$z=0$')
plt.plot(xg[0], h_aniso[0], 'C0--', label='$z=-H$')
plt.plot(xdupuit, hdupuit2, 'C3', label='Dupuit w/ $C$')
plt.legend(loc='lower right');
```

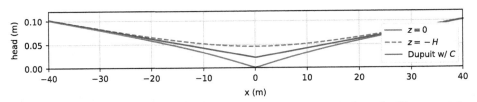

Figure 10.7   Head vs. x at the top and bottom of an anisotropic aquifer with $k_z = k_x/10$ compared to the Dupuit solution with effective conductance C.

The flow field gets more interesting when the inflow from the left is not equal to the inflow from the right ($U_L \neq -U_R$). Three cases are distinguished here; the flow U_L from the left is positive for all three cases. For case 1, water flows to the stream from both sides ($U_R < 0$), which results in a stagnation point along the base of the aquifer. For case 2, the flow on the right side is away from the stream, but not too large ($0 \leq U_R \leq \lambda U_L$). As a result, only part of the water that flows to the stream from the left is captured by the stream and the remainder flows under the stream to the right. There is a stagnation point along the top of the aquifer, and water flows into the stream over the entire stream boundary. For case 3, flow on the right side is away from the stream and larger ($U_R \geq \lambda U_L$). For this case, water flows into the stream over part of the stream boundary and out of the stream along the other part. Examples of all three flow fields are shown in Figure 10.8. The code for the flow nets is the same as that for Figure 10.4 and is not repeated.

(a) 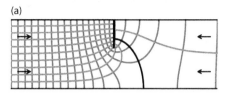 (b)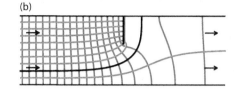

(c)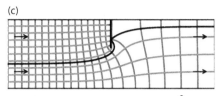

Figure 10.8 Flow nets with $U_L = 0.2$ m²/d for: Case 1 with a stagnation point along the bottom of the aquifer (upper left, $U_R = -0.04$ m²/d); Case 2 with a stagnation point along the top of the aquifer (upper right, $U_R = 0.04$ m²/d); and Case 3 with a stagnation point along the right side of the stream (lower left, $U_R = 0.08$ m²/d).

Exercise 10.1: Plot the head as a function of x at the top and bottom of the aquifer (similar to Figure 10.7) for the three cases of Figure 10.8, and add the Dupuit solution with effective conductance C.

10.3   Flow over a step in the base

The effect of a jump in the base on the flow and head distribution in an unconfined aquifer was investigated in Section 3.2 using a Dupuit solution. Here, the effect of a jump in the base on the flow and head distribution is investigated for confined flow using a solution for two-dimensional flow in the vertical plane. The two-dimensional flow solution is compared to the Dupuit solution, and the effect of a vertically anisotropy hydraulic conductivity is investigated. The solution is obtained with conformal mapping, and here, the function $\zeta(\tau)$ cannot be inverted analytically to get an expression for $\tau(\zeta)$, which complicates the evaluation of the function.

Consider two-dimensional flow in the vertical plane in a confined aquifer with an impermeable top and an impermeable bottom. The base is piecewise horizontal and jumps at $x = 0$ (see Figure 10.9). The aquifer thickness is equal to H_L for $x \leq 0$ and equal to H_R for $x \geq 0$. The hydraulic conductivity is homogeneous and isotropic and equal to k. The flow is equal to $Q_x = U$ from the left to the right

$$\lim_{x \to -\infty} q_x H_L = \lim_{x \to \infty} q_x H_R = U \tag{10.31}$$

The stream function is chosen to equal $\psi = 0$ along the base of the aquifer, which means that $\psi = -U$ along the top of the aquifer. As a reference, the head is equal to zero at the origin

$$h_{x=0, z=0} = 0 \tag{10.32}$$

A solution to the stated problem is again sought in parametric form as $\omega(\tau)$ and $\zeta(\tau)$, where the reference plane τ is chosen as the quarter plane shown in Figure 10.9. Point 1 is chosen to correspond to $\tau = \infty$, point 2 corresponds to $\tau = 0$, and point 4 corresponds to $\tau = 1$. Point 3 corresponds to $\tau = \lambda$, where the value of λ is determined by the mapping. The top of the aquifer and the horizontal portions of the bottom of the aquifer correspond to the positive real axis of τ (note the corresponding red, purple, and brown sections in the ζ and τ planes), while the vertical part of the bottom of the aquifer corresponds to the positive imaginary axis in the τ plane (green).

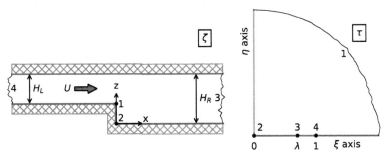

Figure 10.9 Flow over a step in the base; physical ζ plane (left) and reference τ plane (right).

The solution for the stated problem is

$$\zeta = \frac{-H_R}{\pi} \ln \frac{\tau - \lambda}{\tau + \lambda} + \frac{H_L}{\pi} \ln \frac{\tau - 1}{\tau + 1} + (H_R - H_L)i \tag{10.33}$$

$$\omega = \frac{-U}{\pi} \ln[(\tau - 1)(\tau + 1)] + \frac{U}{\pi} \ln[(\tau - \lambda)(\tau + \lambda)/\lambda^2] \tag{10.34}$$

where $\lambda = H_L/H_R$. Note the similarity between Eqs. (10.33) and (10.20), but for this case, as is common in conformal mapping, the solution $\zeta(\tau)$ cannot be inverted to obtain a solution $\tau(\zeta)$. The solutions (10.33) and (10.34) are evaluated by selecting a value of τ and evaluating both the complex potential ω and the corresponding complex coordinate ζ.

A flow net is created as follows. A grid is defined in the τ plane, and both the complex coordinate $\zeta = x + iz$ and the complex potential ω are computed for each grid point. A contour plot is created using the regular `contour` function of `matplotlib` by passing it the computed irregular grid of x and z together with either the computed head or stream function.

All points along the top of the aquifer (between points 3 and 4) correspond to the real axis of τ between $\xi = \lambda$ (point 3) and $\xi = 1$ (point 4). Some points that are far apart along the top of the aquifer in the ζ plane are very close together in the τ plane. This phenomenon is called the crowding effect in conformal mapping. As a result, a regularly spaced grid in the τ plane results in an irregular grid in the ζ plane where some points may be very far apart. An accurate contour plot is obtained by defining an irregularly spaced grid in the τ plane consisting of three segments in the ξ direction, from point 2 to 3, point 3 to 4, and point 4 to a large value; two segments are used in the η plane. Note that neither ζ nor ω can be computed easily at points 3 and 4, as the logarithms in the solution cannot be evaluated at those points. The solution exists in the limit, of course. As a practical solution, the solution is evaluated up to a small distance from points 3 and 4, rather than mathematically taking the limit and programming the result.

An example flow net is shown in Figure 10.10. The streamlines for the Dupuit solution are horizontal lines that jump at the base (compare Figure 3.8) and are shown with dashed lines in Figure 10.10. Note that the Dupuit streamlines are a good approximation of the two-dimensional streamlines beyond one aquifer thickness from the jump in the base. Similarly, the vertical head variation becomes negligible at approximately one aquifer thickness from the jump in the base.

```
# parameters
k = 10 # hydraulic conductivity, m/d
HL = 10 # aquifer thickness on left side, m
HR = 20 # aquifer thickness on right side, m
U = 0.1 # flow from left to right, m^2/d
```

```
# solution
def zetafunc(tau, HR=HR, HL=HL):
    lab = HL / HR
    zeta = -HR / np.pi * np.log((tau - lab) / (tau + lab)) + \
           HL / np.pi * np.log((tau - 1) / (tau + 1)) + \
           (HR - HL) * 1j
    return zeta

def omegafunc(tau, U=U, HR=HR, HL=HL):
    lab = HL / HR
    omega = -U / np.pi * np.log((tau - 1) * (tau + 1)) + \
            U / np.pi * np.log((tau - lab) * (tau + lab) / lab ** 2)
    return omega
```

```
# grid
d = 1e-14 # small offset
lab = HL / HR
xi1 = np.linspace(0, lab - d, 400)
xi2 = np.linspace(lab + d, 1 - d, 400)
xi3 = np.linspace(1 + d, 40, 400)
eta1 = np.linspace(0, 1, 400)
eta2 = np.linspace(1, 40, 400)
xi, eta = np.meshgrid(np.hstack((xi1, xi2, xi3)), np.hstack((eta1, eta2)))
zeta = zetafunc(xi + eta * 1j)
omega = omegafunc(xi + eta * 1j)
```

```
# basic plot
plt.subplot(111, aspect=1, xlim=(-20, 20))
plt.contour(zeta.real, zeta.imag, omega.real / k, np.arange(-0.1, 0.1, 0.002))
plt.contour(zeta.real, zeta.imag, omega.imag, np.arange(-0.1, 0, 0.02))
plt.plot([-40, 0, 0, 40], [10, 10, 0, 0], 'k');
```

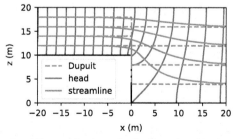

Figure 10.10 Flow net for confined flow with a jump in the base; head contour interval $\Delta h = 2$ mm.

Exercise 10.2: Verify that the conformal mapping solution for ζ (Eq. 10.33) is correct by evaluating the solution at (or very close to) all corner points.

Computation of the complex potential at a specific complex coordinate $\zeta^* = x^* + iz^*$ in the aquifer requires a numerical approach to find the corresponding value of the parameter τ.

For example, τ can be found by minimizing the function

$$f(\tau) = (x^* - x(\tau))^2 + (z^* - z(\tau))^2 \tag{10.35}$$

Minimization can be carried out with a standard nonlinear minimization routine such as the fmin function of scipy.optimize. fmin searches for an array of real variables that minimizes a function, so ξ and η must be entered as an array of length 2 rather than as one complex variable. This approach is implemented below to compute the head at $(x, z) = (-50, 20)$.

```python
def findtau(tauvec, x, z):
    tau = tauvec[0] + tauvec[1] * 1j
    zetastar = zetafunc(tau)
    return ((zetastar.real - x) ** 2 + (zetastar.imag - z) ** 2)

from scipy.optimize import fmin
xi0, eta0 = fmin(findtau, (1, 1), args=(-50, 20), disp=0)
tau0 = xi0 + eta0 * 1j
print(f'tau: {tau0}')
print(f'zeta: {zetafunc(tau0)}')
```

```
tau: (0.9999999664789975+1.245910302309445e-11j)
zeta: (-49.99698501153817+19.99881690414834j)
```

Next, two examples are presented for the case of a vertically anisotropic hydraulic conductivity. In the first example, the vertical anisotropy is the popular value of $k_z = k_x/10$. The corresponding flow net is shown in Figure 10.11. It can be seen from the flow net that it takes about three ($\sim \sqrt{k_x/k_z}$) aquifer thicknesses downstream of the jump in the base before the streamlines of the Dupuit solution are a reasonable approximation of the streamlines of the two-dimensional solution. At the jump in the base, the difference between the head at the top of the aquifer and the head at the bottom of the jump is 2 cm.

```python
# additional parameters anisotropic case 1
kx = 10 # horizontal hydraulic conductivity, m/d
kz = 1 # vertical hydraulic conductivity, m/d
```

```python
# solution anisotropic case
ktilde = np.sqrt(kx * kz)
zetaaniso = zetafunc(xi + eta * 1j)
omeganiso = omegafunc(xi + eta * 1j)
xg = zetaaniso.real / np.sqrt(kz / kx)
zg = zetaaniso.imag
h = omeganiso.real / ktilde
psi = omeganiso.imag
```

```python
# basic plot anisotropic case
plt.subplot(111, aspect=1, xlim=(-20, 60))
plt.contour(xg, zg, h, np.arange(-0.1, 0.1, 0.002))
plt.contour(xg, zg, psi, np.arange(-0.1, 0, 0.02))
plt.plot([-40, 0, 0, 40], [10, 10, 0, 0], 'k');
```

The second example of vertical anisotropy is for the case that the vertical hydraulic conductivity is 100 times larger than the horizontal hydraulic conductivity. The much larger vertical than horizontal hydraulic conductivity approaches the Dupuit approximation (in which the vertical resistance to flow is completely neglected). The two solutions should thus be very similar. The flow net is shown in Figure 10.12; the code to produce the flow net is the same

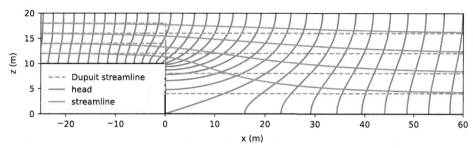

Figure 10.11 Flow net for confined flow with a jump in the base and vertical anisotropy $k_x/k_z = 10$; head contour interval $\Delta h = 2$ mm.

as that for Figure 10.11 and is not repeated here. The head contours are indeed almost vertical, even near the jump in base, and the streamlines are horizontal and almost jump down vertically at the jump in the base.

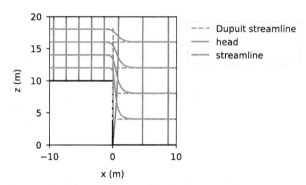

Figure 10.12 Flow net for confined flow with a jump in the base and vertical anisotropy $k_z = 100k_x$, which approaches the Dupuit approximation; head contour interval $\Delta h = 2$ mm.

Exercise 10.3: Compute the head drop between $(x, z) = (-50, 20)$ and $(x, z) = (50, 20)$ for the two-dimensional isotropic solution.

Exercise 10.4: Compute the head drop between $x = -50$ and $x = 50$ for the Dupuit solution, and compute the additional head drop across the jump for the two-dimensional solution (previous exercise) as compared to the Dupuit solution for the isotropic case.

10.4 Spatially varying head at the top of the aquifer

The water table in an unconfined aquifer is sometimes very roughly approximated as a subdued replica of the topographic surface. This makes some intuitive sense, because groundwater often discharges at a low point in the landscape. Groundwater that recharges below the topographic highs has to take a relatively long path to the discharge area in the topographic lows, which necessitates a hydraulic head gradient, and thus a mounding of the water table. It turns out, however, that the water table and the topographic surface are poorly correlated or even uncorrelated in many unconfined aquifers (Haitjema and Mitchell-Bruker, 2005). Nonetheless, an undulating water table may be a useful approximation to conceptualize regional flow patterns.

In this section, flow systems are investigated where the head at the top of the aquifer is specified to consist of one or more sinusoidal functions, approximating a subdued sequence of topographic highs and lows. The problem is discussed here because it resembles the famous solution of Toth (1963), which is often used to investigate nested flow domains in basin-scale models. As this is a problem where the head is specified at the top of the aquifer, it must be verified that the resulting flow through the top of the aquifer is hydrologically realistic.

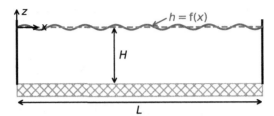

Figure 10.13 Vertical cross section of an aquifer where flow is driven by a specified head at the top of the aquifer.

Consider steady two-dimensional flow in the vertical plane (Figure 10.13). The hydraulic conductivity is k, and the bottom of the aquifer is horizontal at $z = -H$. The aquifer is a distance L long, and the thickness of the aquifer is approximated as constant and equal to H. The flow is governed by Laplace's equation (10.4). The lateral boundaries and the base of the aquifer are impermeable

$$q_x|_{x=0,z} = q_x|_{x=L,z} = q_z|_{x,z=-H} = 0 \tag{10.36}$$

The head h at the top of the aquifer is specified to vary periodically as a cosine with amplitude a and period $p = 2L/n$, where n is an arbitrary integer

$$h|_{x,z=0} = a\cos(2\pi x/p) \tag{10.37}$$

In terms of the specific discharge potential, the boundary conditions are

$$\left.\frac{\partial\phi}{\partial x}\right|_{x=0,z} = \left.\frac{\partial\phi}{\partial x}\right|_{x=L,z} = \left.\frac{\partial\phi}{\partial z}\right|_{x,z=-H} = 0 \tag{10.38}$$

and

$$\phi|_{x,z=0} = A\cos(2\pi x/p) \tag{10.39}$$

where $A = ka$.

The solution for the specific discharge potential is obtained from separation of variables as (e.g., Craig, 2008)

$$\phi = \frac{A\cos(2\pi x/p)\cosh[2\pi(z+H)/p]}{\cosh(2\pi H/p)} \tag{10.40}$$

The corresponding complex potential is

$$\omega = A\frac{\cos[2\pi(\zeta + Hi)/p]}{\cosh(2\pi H/p)} \tag{10.41}$$

where $\zeta = x + iz$. The complex specific discharge is obtained as

$$q_x - iq_z = -\frac{d\omega}{d\zeta} = \frac{2\pi A}{p}\frac{\sin[2\pi(\zeta + Hi)/p]}{\cosh(2\pi H/p)} \tag{10.42}$$

The vertical flux at the top of the aquifer is a cosine function, just like the head

$$q_z = \frac{-2\pi A}{p}\frac{\cos(2\pi x/p)\sinh[2\pi(z+H)/p]}{\cosh(2\pi H/p)} \tag{10.43}$$

As a first example, consider the case where the head at the top of the aquifer is a sequence of two highs and two lows, starting with a high of 0.2 m on the left and ending with a low of −0.2 m on the right. The head specified at the top of the aquifer, the resulting vertical flux at the top of the aquifer, and the resulting flow net are plotted in Figure 10.14. The resulting flow field has three local flow cells: Water flows from the highs to the lows. All three flow cells are identical. As the head is specified at the top of the aquifer, it must be checked whether the vertical flux at the top of the aquifer (i.e., the recharge) is hydrologically possible. For the given parameters, the highest vertical flux at the top of the aquifer is ∼ 1.5 mm/d, which can be possible from precipitation in many climates. The flow net for the same situation, but with a ten times higher hydraulic conductivity, looks exactly the same, but the highest vertical flux at the top of the aquifer is also 10 times higher. This requires a recharge of 15 mm/d, which seems unrealistic except maybe for the wettest regions on the planet.

```
# parameters
k = 1 # hydraulic conductivity, m/d
L = 1200 # length of aquifer, m
H = 200 # thickness of aquifer, m
a = 0.2 # amplitude of head variation, m
p = 2 * L / 3 # period of head variation, m
A = k * a # amplitude of potential variation, m^2/d
```

```
# solution
def omega(x, z, H, A, p):
    zeta = x + 1j * z
    return A * np.cos(2 * np.pi * (zeta + H * 1j) / p) / \
                np.cosh(2 * np.pi * H / p)

def discomplex(x, z, H, A, p):
    zeta = x + 1j * z
    return 2 * np.pi * A / p * np.sin(2 * np.pi * (zeta + H * 1j) / p) / \
                        np.cosh(2 * np.pi * H / p)

xg, zg = np.meshgrid(np.linspace(0, L, 100), np.linspace(-H, 0, 100))
om = omega(xg, zg, H, A, p)
wdis = discomplex(xg, zg, H, A, p)
htop = om.real[-1] / k
qtop = -wdis.imag[-1]
```

```
# basic plot h, qz, flow net, sharing the x-axis
fig, ax = plt.subplots(3, 1, sharex=True, squeeze=True,
                        gridspec_kw={'height_ratios':[1, 1, 2]})
plt.sca(ax[0])
plt.plot(xg[0], htop)
plt.sca(ax[1])
plt.plot(xg[0], qtop)
plt.sca(ax[2])
plt.contour(xg, zg, om.real / k, np.arange(-0.2, 0.21, 0.02))
plt.contour(xg, zg, om.imag, np.arange(-0.2, 0.21, 0.02))
plt.axis('scaled');
```

The flow in one flow cell is compared to the flow obtained with a Dupuit model for the same situation. The Dupuit solution for the head is simple, as it is equal to the boundary condition at the top of the aquifer (Eq. 10.37)

$$h = a\cos(2\pi x/p) \tag{10.44}$$

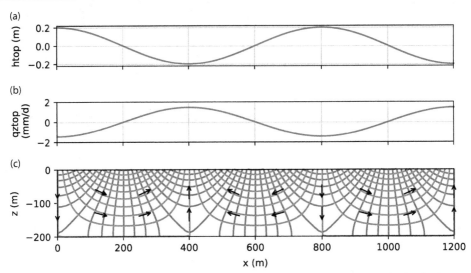

Figure 10.14 Flow caused by periodically varying specified head at the top of the aquifer (top), the resulting vertical flow at the top of the aquifer (middle), and the resulting flow net with $\Delta h = 2$ cm and $\Delta \psi = 0.02$ m^2/d (bottom).

The corresponding expression for the Dupuit discharge vector is

$$Q_x = \frac{kHa2\pi}{p} \sin(2\pi x/p) \tag{10.45}$$

For the two-dimensional solution, the vertically integrated flow in the aquifer is equal to minus the stream function at the top of the aquifer (as $\psi = 0$ along the bottom). The two-dimensional solution is compared to the Dupuit solution for the case that there is one flow cell in the 1200-meter-long domain; the Dupuit solution is a reasonable approximation for a flow cell that is 1200 m long and 200 m deep (Figure 10.15). For the case that there are three flow cells in the domain (as in Figure 10.14), the Dupuit solution is a poor approximation as one flow cell is 400 m long and 200 m deep, which is far from predominantly horizontal.

```
# solution for flow cell of 1200 m
L = 1200
p = 2 * L # one flow cell
x = np.linspace(0, L, 100)
Qx2D = -omega(x, 0, H, A, p).imag # Qx2D = -psi
QxDupuit = k * H * a * 2 * np.pi / p * np.sin(2 * np.pi * x / p)
# grid of stream function
xg, zg = np.meshgrid(np.linspace(0, L, 400), np.linspace(-H, 0, 100))
psi2D = omega(xg, zg, H, A, p).imag
xd = xg[0]
zd = [-H, 0]
psiDupuit = np.zeros((2, len(xd)))
psiDupuit[1] = -k * H * a * 2 * np.pi / p * np.sin(2 * np.pi * xd / p)
```

```
# basic plot
plt.subplot(211)
plt.plot(x, Qx2D, label='2D solution')
plt.plot(x, QxDupuit, label='Dupuit solution')
plt.legend()
```

```
plt.subplot(212, aspect=1)
plt.contour(xg, zg, psi2D, np.arange(-0.2, 0, 0.02), colors='C1')
plt.contour(xd, zd, psiDupuit, np.arange(-0.2, 0, 0.02), colors='C0');
```

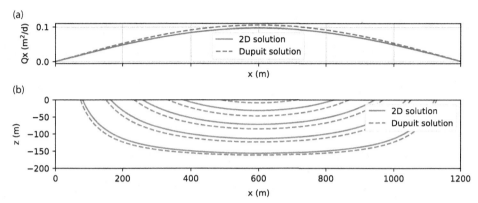

Figure 10.15 Comparison of two-dimensional flow solution to Dupuit solution for one flow cell of 1200 m length; Q_x (top) and streamlines (bottom).

The head fluctuation at the top of the aquifer can be an arbitrary combination of cosine functions (Eq. 10.37) with different amplitudes and different periods. For example, it is possible to approximate the head at the top of the aquifer with a constant slope $-\Delta h/L$ by a Fourier series plus a constant

$$-\Delta h \frac{x}{L} \approx -\frac{\Delta h}{2} + \Delta h \sum_{n=1,3,5,\ldots}^{\infty} \frac{4}{n^2 \pi^2} \cos\left(\frac{n\pi x}{L}\right) \qquad (10.46)$$

As an example, the flow problem is solved for the case that the head at the top of the aquifer consists of a constant slope plus one periodic variation

$$h|_{x,z=0} = -\Delta h \frac{x}{L} + a \cos(2\pi x/p) \qquad (10.47)$$

The resulting flow net (Figure 10.16) consists of three local flow cells and one regional flow cell where water that infiltrates on the left exfiltrates all the way to the right of the aquifer, flowing below the local flow cells. The regional flow cell only occurs when the aquifer is deep enough. Note that it is again questionable whether the resulting vertical flux at the top of the aquifer is hydrologically possible for this case, as is explored in the exercises.

```
# additional parameters and changed parameters
delh = 2 # head drop over distance L, m
p = L / 3
# period of head variation, m
nterms = 100 # number of terms in Fourier series
```

```
# solution
xg, zg = np.meshgrid(np.linspace(0, L, 200), np.linspace(-H, 0, 200))
om = np.zeros(xg.shape, dtype='complex')
om -= 0.5 * delh
for n in range(1, nterms + 1, 2):
    A = 4 * k * delh / (n ** 2 * np.pi ** 2)
    om += omega(xg, zg, H, A, 2 * L / n)
```

```
om += omega(xg, zg, H, k * a, p)
htop = om.real[-1] / k
```

```
# basic flow net
plt.subplot(111, aspect=1)
plt.contour(xg, zg, om.real, 10, colors='C0')
plt.contour(xg, zg, om.imag, 10, colors='C1');
```

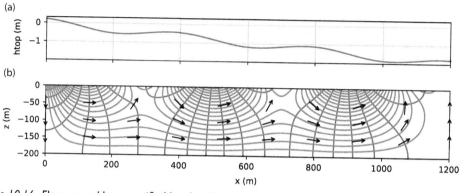

Figure 10.16 Flow caused by a specified head at the top of the aquifer consisting of a constant slope and a periodically varying specified head; specified head (top) and resulting flow net (bottom).

Exercise 10.5: Compute and plot q_z at the top of the aquifer for the example of Figure 10.16.

Exercise 10.6: Create flow nets for three different aquifer thicknesses: $H = 70$ m, $H = 90$ m, and $H = 110$ m, and determine for which of these aquifer thicknesses there are no regional flow cells.

10.5   Interface flow toward the coast

Several solutions for interface flow in coastal aquifers were discussed in this book so far (Chapter 4 and Section 7.5), all based on the Dupuit approximation. The width of the outflow zone at the coast was zero for all presented Dupuit solutions, except for the solution with a resistance layer along the sea bottom (Section 4.3). In Chapter 4, it was claimed that the resulting interface was, nevertheless, a good approximation, even though the flow near the coast has a significant vertical component. In this and the following section, Dupuit interface solutions are compared to solutions for two-dimensional flow in the vertical plane, and it is investigated how wide the outflow zone below the sea is.

Consider steady two-dimensional interface flow in the vertical plane in a confined coastal aquifer. The aquifer is so deep that the interface does not reach the bottom of the aquifer; the saltwater is at rest (Figure 10.17). The density of the freshwater is ρ_f, the density of the saltwater is ρ_s, and the hydraulic conductivity of the aquifer is isotropic and equal to k. The sea bottom is horizontal and at the same elevation as the top of the confined aquifer. The origin of the coordinate system is chosen at the coastline at the top of the aquifer. The flow toward the coast is constant and equal to $Q_x = U$. The stream function along the top of the aquifer is chosen to equal $\psi = 0$ so that the stream function along the interface equals $\psi = U$.

This problem is known as a free boundary problem. The location of the interface is a free boundary because it is unknown prior to solving the problem. There are two conditions specified along the interface: The interface is an impermeable boundary, and the elevation of the interface z_i is related to the head h_i along the interface according to the Ghyben–Herzberg equation (Chapter 4)

$$z_i = -\alpha h_i \tag{10.48}$$

where

$$\alpha = \frac{\rho_f}{\rho_s - \rho_f} \tag{10.49}$$

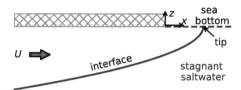

Figure 10.17   Interface flow toward the coast.

The stated problem is known as Glover's problem (Glover, 1959). The complex potential solution used here is obtained from Bruggeman (1999, Eq. 910.01) as

$$\omega = \sqrt{-2kU\zeta/\alpha} \tag{10.50}$$

where $\zeta = x + iz$ is the complex coordinate. The solution for the complex potential can be inverted, which gives

$$\zeta = \frac{-\omega^2 \alpha}{2kU} \tag{10.51}$$

The specific discharge potential along the interface equals $\phi_i = -kz_i/\alpha$, and the stream function is equal to U. An equation for the elevation of the interface as a function of x may be obtained by substituting $\omega = -k\alpha z_i + Ui$ for ω and $x + iz_i$ for ζ in Eq. (10.50), which gives, after some algebra,

$$z_i = -\sqrt{\frac{U^2\alpha^2}{k^2} - \frac{2U\alpha x}{k}} \tag{10.52}$$

The tip of the interface is the point where the interface intersects the sea bottom (Figure 10.17). The location of the tip of the interface x_{tip} is obtained by setting $z_i = 0$ in Eq. (10.52) and solving for x, which gives

$$x_{\text{tip}} = \frac{U\alpha}{2k} \tag{10.53}$$

This means that the outflow zone is wider for a larger value of the flow toward the coast U.

In the example below, a flow net is computed by specifically computing the locations of the streamlines and equipotentials, as ζ is known as a function of ω (Eq. 10.51). For the chosen parameters, the interface is already 7 m deep at a distance of 14 m inland from the coastline (Figure 10.18). The length of the outflow face is just 0.8 m.

```
# parameters
k = 10 # hydraulic conductivity, m/d
rhof = 1000 # density freshwater, kg/m^3
```

```
rhos = 1025 # density of saltwater, kg/m^3
U = 0.4 # flow toward the coast, m^2/d
```

```
# solution
alpha = rhof / (rhos - rhof)
n = 100
zeta_streamlines = []
zeta_equipotentials = []
for psi in np.linspace(0, U, 6):
    omega = np.linspace(0, 2, n) + 1j * psi
    zeta = -omega**2 * alpha / (2 * k * U)
    zeta_streamlines.append(zeta)
for phi in np.arange(0, 2, U / 6):
    omega = phi + 1j * np.linspace(0, U, n)
    zeta = -omega**2 * alpha / (2 * k * U)
    zeta_equipotentials.append(zeta)
zetatip = -(1j * U) ** 2 * alpha / (2 * k * U)
print(f'location of the tip: {zetatip.real: .2f} m')
```

```
location of the tip:  0.80 m
```

```
# basic plot
plt.subplot(111, aspect=1, xlim=(-15, 1))
for zeta in zeta_equipotentials:
    plt.plot(zeta.real, zeta.imag, 'C0')
for zeta in zeta_streamlines:
    plt.plot(zeta.real, zeta.imag, 'C1')
```

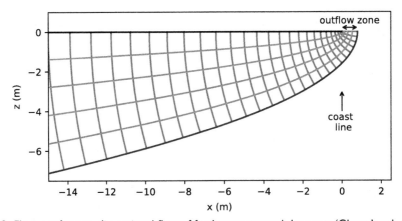

Figure 10.18 Flow net for two-dimensional flow of freshwater toward the coast (Glover's solution).

The interface position obtained with Glover's solution is compared to the Dupuit interface solution. The Dupuit solution for the elevation of the interface is obtained from Section 4.1 as

$$z_i = -\sqrt{\frac{-2U\alpha x}{k}} \qquad (10.54)$$

Note that z_i^2 of the two-dimensional solution (10.52) differs by a constant factor from the Dupuit solution. The difference between the Glover solution and the Dupuit solution is minor for the parameters of the example (Figure 10.19).

The difference between the two solutions becomes a bit larger when the vertical hydraulic conductivity k_z of the aquifer is significantly smaller than the horizontal hydraulic conductivity k_x. The Glover solution is modified for an anisotropic hydraulic conductivity using the approach of Section 10.1. The solution for the complex potential is

$$\tilde{\omega} = \sqrt{-2\tilde{k}U\tilde{\zeta}/\alpha} \tag{10.55}$$

The corresponding elevation of the interface is

$$z_i = -\sqrt{\frac{U^2\alpha^2}{k_x k_z} - \frac{2U\alpha x}{k_x}} \tag{10.56}$$

and the location of the tip of the interface is

$$x_{\text{tip}} = \frac{U\alpha}{2k_z} \tag{10.57}$$

In the example below, the interface is plotted for the isotropic case and for the case that $k_z = k_x/10$, which results in an outflow zone that is 10 times wider than for the isotropic case (Figure 10.19). The interface is located somewhat deeper than for the isotropic case. The Dupuit solution does not change, of course, as the resistance to flow in the vertical direction is neglected.

The Dupuit solution is a very good approximation of the isotropic Glover solution and a reasonable approximation of the anisotropic Glover solution. The Dupuit solution is always conservative when assessing the extent of seawater intrusion: The thickness of the freshwater zone is always smaller and the seawater intrusion is more severe than in the Glover solution. The largest difference in the thickness of the freshwater zone is at the coastline.

```
# additional parameters anisotropic case
kx = k
kz = k / 10
```

```
# isotropic solution
xtip = U * alpha / (2 * k)
xglover = np.linspace(-150, xtip, 100)
zglover = -np.sqrt(U * alpha / k * (U * alpha / k - 2 * xglover))
# anisotropic solution
xtipaniso = U * alpha / (2 * kz)
xaniso = np.linspace(-150, xtipaniso, 200)
zaniso = -np.sqrt(U * alpha / kx * (U * alpha / kz - 2 * xaniso))
print(f'location of the tip for anisotropic case: {xtipaniso: 0.2f} m')
#
xdupuit = np.linspace(-150, 0, 100)
zdupuit = -np.sqrt(-2 * U * alpha * xdupuit / k)
```

```
location of the tip for anisotropic case:  8.00 m
```

```
# basic plot
plt.plot(xglover, zglover, 'C1', label='Glover $k_z=k_x$')
plt.plot(xaniso, zaniso, 'C2', label='Glover $k_z=k_x/10$')
plt.plot(xdupuit, zdupuit, 'C1--', label='Dupuit $H=20$')
plt.legend();
```

Exercise 10.7: Compute the ratio of the depth of the interface to the head at the interface at $x = -50$ and at $x = -100$ for the anisotropic case. (Hint: The answer is, of course, α if you do this correctly.) Next, compute the ratio of the depth of the interface to the head at the top of the aquifer at $x = -50$ and at $x = -100$ for the anisotropic case.

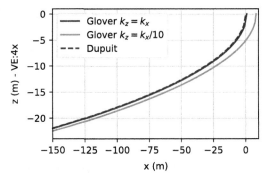

Figure 10.19 Comparison of interface elevation between the Glover solution (both isotropic and anisotropic) and the Dupuit solution.

10.6   Interface flow below a strip island

Two-dimensional steady interface flow in the vertical plane is further explored by considering interface flow in a cross section of a strip island; the saltwater is at rest. The distance between the two shores (i.e., the width of the island) is $2L$, and the infiltration rate is equal to N (Figure 10.20). The flow is unconfined, but the small variation in the phreatic surface is neglected and the top of the aquifer is approximated as horizontal at the same elevation as the sea bottom. The origin of the coordinate system is chosen at the top of the aquifer in the center of the island. The sea bottom is horizontal, and the head along the sea bottom is set equal to zero

$$h|_{x>L,z=0} = h|_{x<-L,z=0} = 0 \tag{10.58}$$

The stream function along the interface is set equal to zero.

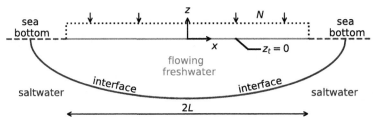

Figure 10.20 Two-dimensional interface flow in a vertical cross section of a strip island.

The complex specific discharge potential solution for the stated problem is obtained from Bruggeman (1999, Eq. 910.03), modified for the coordinate system chosen here. The solution has the following form

$$\omega^2 + 2iN\zeta\omega + kN\zeta^2/\alpha - N(N + k/\alpha)L^2 = 0 \tag{10.59}$$

Note that Eq. (10.59) is a second-order polynomial in ω. A solution for ω for a given value of ζ is obtained using the standard formula for the root of a parabola, which gives

$$\omega = -iN\zeta + \sqrt{N^2(L^2 - \zeta^2) + Nk(L^2 - \zeta^2)/\alpha} \tag{10.60}$$

The specific discharge potential along the interface equals $\phi_i = -kz_i\alpha$ so that an equation for the elevation of the interface as a function of x may be obtained by substituting $\omega = -kz_i\alpha$

for ω and $x + iz_i$ for ζ in Eq. (10.59), which gives (after considerable algebra)

$$z_i = -\sqrt{\frac{N\alpha}{k}L^2 - \frac{N\alpha}{N\alpha + k}x^2} \tag{10.61}$$

The shape of the interface is an ellipse, and the locations of the shorelines are the foci of the ellipse. The location x_{tip} of the tip of the interface is obtained by setting $z_i = 0$ in the previous equation, which gives

$$x_{\text{tip}} = \pm L\sqrt{1 + \frac{N\alpha}{k}} \tag{10.62}$$

The elevation of the interface at the center of the island is obtained by setting $x = 0$ in Eq. (10.61), which gives

$$z_i|_{x=0} = -L\sqrt{\frac{N\alpha}{k}} \tag{10.63}$$

A flow net is constructed as follows. First, the location of the tip is computed with Eq. (10.62). Second, an array of x values is created from $-x_{\text{tip}} + \varepsilon$ to $+x_{\text{tip}} - \varepsilon$, where ε is a small number (ε is used to make sure that the first point is numerically not to the left of $-x_{\text{tip}}$ and the last point is numerically not to the right of $+x_{\text{tip}}$). Third, a grid of z values is computed between the interface and the top of the aquifer for every x value. Fourth and final, ω is computed on the grid and a flow net is contoured.

In the example below, the width of the island equals 2000 m; the flow net is shown in Figure 10.21. The freshwater zone is a bit more than 63 m thick at the center of the island, while the outflow zone is only 2 m on each side. Note that, contrary to other flow nets in this chapter, the vertical scale is exaggerated in Figure 10.21 (otherwise, the figure would be unreadable), which makes the flow net resemble flow in an anisotropic aquifer while this is not the case.

```
# parameters
k = 10 # hydraulic conductivity, m/d
L = 1000 # half-width of island, m
N = 0.001 # areal recharge, m/d
rhof = 1000 # densityof freshwater, kg/m^3
rhos = 1025 # density of saltwater, kg/m^3
alpha = rhof / (rhos - rhof)
```

```
# solution
def omega(zeta, k=k, alpha=alpha, L=L, N=N):
    om = -1j * N * zeta + np.sqrt(N ** 2 * (L ** 2 - zeta ** 2) +
                          N * k * (L ** 2 - zeta ** 2) / alpha)
    return om
```

```
# compute interface
ng = 201
xtip = L * np.sqrt(1 + N * alpha / k)
print(f'length outflow zone: {xtip - L:0.2f} m')
xi = np.linspace(-xtip + 1e-12, xtip - 1e-12, ng)
zi = -np.sqrt(N * alpha / k * L ** 2 - N * alpha / (N * alpha + k) * xi ** 2)
zcenter = -L * np.sqrt(N * alpha / k)
print(f'elevation of interface at center of island: {zcenter:0.2f} m')
# compute grid for contouring
```

```
xg = xi * np.ones((50, ng))
zg = np.zeros_like(xg)
for i in range(ng):
    zg[:, i] = np.linspace(zi[i], -1e-12, 50)
om = omega(xg + 1j * zg)
```

length outflow zone: 2.00 m
elevation of interface at center of island: -63.25 m

```
# basic plot
plt.subplot(111, aspect=5, xlim=(-xtip, xtip))
plt.contour(xg, zg, om.imag, 20)
plt.contour(xg, zg, om.real, 20)
plt.plot(xi, zi, 'k');
```

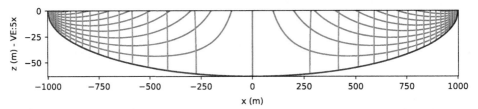

Figure 10.21   Flow net for two-dimensional interface flow in a cross section of a strip island with isotropic hydraulic conductivity (note the vertical exaggeration VE = 5×).

The solution for an anisotropic hydraulic conductivity is obtained using the approach of Section 10.1. In addition, the infiltration rate $\tilde{N}$ in the transformed domain is

$$\tilde{N} = N\sqrt{k_x/k_z} \tag{10.64}$$

The resulting equation for the elevation of the interface is

$$z_i = -\sqrt{\frac{N\alpha}{k_x}L^2 - \frac{N\alpha}{N\alpha + k_z}\frac{k_z}{k_x}x^2} \tag{10.65}$$

Hence, the elevation of the interface at the center of the island is

$$z_i|_{x=0} = -\sqrt{\frac{N\alpha}{k_x}L^2} \tag{10.66}$$

This result may be surprising, as it means that the depth of the interface at the center of the island only depends on the horizontal component k_x of the hydraulic conductivity and not on the vertical component. Conversely, the equation for the tip of the interface only depends on k_z and not on k_x

$$x_{\text{tip}} = L\sqrt{1 + \frac{N\alpha}{k_z}} \tag{10.67}$$

This means that the length of the outflow zone is larger when the vertical hydraulic conductivity is smaller.

The solution for two-dimensional flow in the vertical plane is compared to the Dupuit solution. The Dupuit solution for the interface is obtained from Section 4.2 as

$$z_i = -\sqrt{\frac{N\alpha}{k_x}(L^2 - x^2)} \tag{10.68}$$

so that the Dupuit elevation of the interface at the center of the island is

$$z_i|_{x=0} = -\sqrt{\frac{N\alpha}{k_x}L^2}$$

(10.69)

which is exactly the same as the two-dimensional solution (Eq. 10.66) irrespective of the vertical hydraulic conductivity! Everywhere else, the Dupuit solution is conservative: The thickness of the freshwater zone is smaller than in the two-dimensional solution.

In the example below, the interface obtained with the Dupuit solution is compared to the two-dimensional solution for the case that the hydraulic conductivity is isotropic and for the case that the hydraulic conductivity is anisotropic. The Dupuit solution is again a very good and conservative approximation (Figure 10.22).

```
# additional parameters for anisotropic case
kx = k
kz = kx / 20
```

```
# solution
xtipaniso = L * np.sqrt(1 + N * alpha / kz)
xaniso = np.linspace(-xtipaniso + 1e-12, xtipaniso - 1e-12, 400)
zianiso = -np.sqrt(N * alpha / kx * L ** 2 -
                  N * alpha / (N * alpha + kz) * kz / kx * xaniso ** 2)
xdupuit = np.linspace(-L, L, 200)
zidupuit = -np.sqrt(N * alpha / kx * (L ** 2 - xdupuit ** 2))
```

```
# basic plot
plt.subplot(111, aspect=5)
plt.plot(xi, zi, label='2D, $k_z=k_x$')
plt.plot(xdupuit, zidupuit, '--', label='Dupuit')
plt.plot(xaniso, zianiso, label='2D, $k_z=k_x/20$')
plt.legend();
```

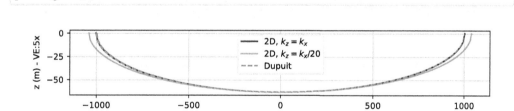

Figure 10.22 Comparison of the interface position below a strip island for the two-dimensional solution and the Dupuit solution.

Exercise 10.8: Compute the ratio of the depth of the interface to the head at the interface at $x = 0$ for the anisotropic case. Next, compute the ratio of the depth of the interface to the head at the top of the aquifer at $x = 0$ for the anisotropic case.

Python primer

The aim of this primer is to provide an introduction to Python's computational capabilities. This primer is not intended to be a general introduction to computer programming, but a basic introduction to understand the code examples in this book, geared toward novice Python users. Readers already skilled at programming in Python will probably choose to skip this appendix, although some parts may be useful as a refresher.

Python is an open-source, object-oriented programming language used for both stand-alone programs and scripting applications. It was initially developed by Guido van Rossum, who named it after Monty Python's Flying Circus. Over the past decades, Python has become extremely popular as a programming language in science and engineering. This is not only because it is one of the easiest programming languages to learn, but also because it has many extensions that make Python a programming language that can be used for an enormous range of applications. Other advantages are that it is open source and runs on all major operating systems.

Python consists of the core language and comes with a comprehensive standard library. In addition, there are many third-party packages (or libraries) for tasks that are not included in the standard library. Both Python and third-party packages are being developed and updated regularly. The code in this book is written in Python 3.8, but it is expected that most, if not all, examples will work with newer versions.

Python is an interpreted language, which means that it is not necessary to compile it into binary machine code before a program can be run. Instead, the commands are typed (or read from a script) in an interpreter, which may be embedded in a web browser, as is the case when using Jupyter Notebooks.

Users that want to run the code examples in this book on their own computer must first install Python and all the required extensions and tools. In principle, one can download the core programming language and standard library from python.org and then download and install everything else that is required, but it is much better and easier to download one of the so-called Python distributions, which includes Python and a whole suite of extensions that are pre-packaged into a single installer. The Anaconda distribution is the recommended distribution to use because it is very comprehensive. All packages used in the code examples in this book (and many more) are included in the Anaconda distribution. Another advantage of Anaconda is that it is easy to update packages and that Anaconda usually installs without major hassles. To install Anaconda, go to the web site www.anaconda.com/products/individual and download the installer for your operating system. Once downloaded, double-click on the installer and choose next or I agree a couple of times until Anaconda is installed (unless you have a very strict system administrator).

Jupyter Notebooks (or simply "notebooks") provide a way to execute Python code in a web browser. A notebook consists of cells, which either have text or code. The code cells can easily be recognized as they are preceded by In [ ]: or In [n]: where n is a counter. To see what a code fragment does, position your cursor in one of the code cells of a notebook (either by clicking the left mouse button or by navigating using the up and down arrow keys), hold the [shift] key, and hit [enter] or click the Run button. The output, preceded by Out [n]: (where n is a number), then appears below the code cell.

Each time one of the code cells in a notebook gets executed, n gets increased by 1. This means that if code cells are executed one by one starting from the top, they will be numbered successively from top to bottom. But if cells are not executed in a systematic way, which may happen when a user is experimenting with the code examples, the numbering may differ. More often than not, a user does not have to worry about the numbering, but if there are any problems with the code, it may help to remember that the last cell that was executed has the highest number n.

It is possible to change the code cells to play around with the examples, but note that some examples later on in a notebook may not work as intended anymore. If a code cell behaves unexpectedly, it may be because something was defined erroneously in one of the cells above it. To help locate the problem, clear all the code output from the notebook by selecting Kernel → Restart Kernel and Clear All Outputs from the menu and start executing again from the top.

A.1   Basics

There are a few things that are worth mentioning upfront about Python, as they form typical traps for novice users. First of all, Python is case sensitive, so a variable with the name avariable is not the same as Avariable or AVARIABLE. Second, systematic indentation is not only good coding style, but it is a feature of the Python language. Third, Python starts counting at zero (for those who have trouble getting used to this, it may help to remember that the first hour of a day also starts at zero, or that you don't turn one until one year after you were born). All three aspects will become clear in the examples that follow.

Because Python is an interpreted language, one of its uses is as a basic calculator. For example, to compute 6 * 12, type

```
6 * 2
```

12

The spaces are added around the multiplication symbol to make the code more readable. 6*2 works just as well, but it is not as easy to read and is considered to be poor style. Style guidelines for Python can be found at www.python.org/dev/peps/pep-0008/. The example above also shows how single-line comments can be included by typing text behind a # symbol. Python ignores any code on a line after a # symbol.

Exponentiation is performed by using **

```
2 ** 6
```

64

A hyphen (minus sign) swaps the sign of a number. Some care must be taken to make sure that the intended numerical operations are carried out in the correct order. In the following example, the number 2 is first raised to the power of 4, and only then is the sign swapped

```
-2 ** 4
```

-16

If, instead, the desired outcome is to raise −2 to the power 4, parentheses are required

```
(-2) ** 4
```

16

Values may be stored in variables by typing

```
a = 6
b = 2
a * b
```

12

Both a and b are now variables. Once a variable has been created in a session, it remains in memory, so it can be used in other code cells. Arithmetic operations can be performed on variables in the same way as on numbers. Because various operations can be performed on variables, their numerical value can change during the course of a program.

Variable names cannot have spaces, nor can they start with a number. As mentioned earlier, variable names are case sensitive, so a variable called v is not the same as V. Some words are reserved and can't be used as variable names, such as and, True, False, and lambda. Care must be taken not to choose variable names that have the same name as existing Python functions. Many code editors, including Jupyter Notebooks, recognize reserved words and color-code them (called syntax highlighting), which helps to avoid mistakes.

Each variable has a type. The most common variable types include the following:

1   Integer
2   Floating-point number (or 'float')
3   String (which contains text)
4   Boolean (which is either true or false)
5   Complex number

Unlike some other programming languages, there is no need to explicitly specify the type of a variable when it is first defined because Python infers it automatically, as demonstrated in the following example

```
a1 = 6 # integer
a2 = 6. # float
a3 = 'a string is enclosed by single or double quotes' # string
a4 = True # boolean
a5 = 2 + 3j # complex number
```

Note that the imaginary part of a complex number is indicated by j (rather than i). Also note that executing this code cell in a notebook does not generate any output, as was the case with the previous code cells. Only the last line of a code cell automatically prints output to the screen, except when the output is stored in a variable or when a semi-colon is added at the end of the last line.

In general, an operation on two variables of the same type results in a variable of the same type. For example, multiplication of two integers results in an integer

```
8 * 4 # results in an integer
```

32

The only exception is that the division of two integers automatically gives a float, even if there is no remainder. (Historical note: This behavior used to be different and quite confusing in Python 2.)

```
8 / 4 # results in a float
```

2.0

Before moving on, it might be useful to discuss the help options in a notebook. Help for functions can be obtained in several ways. The easiest is to type the name of the function in a code cell and then open a parenthesis like (if code-completion is active, a closing parenthesis might appear automatically)

```
float(
```

and then hit the [shift]+[tab] keys together to pop-up a help box. There is a little + in the upper right corner of the help box. When you click on the +, the box expands, and you can scroll through all the help. Hit the x in the upper-right corner to close the help box. Alternatively, you can type

```
float?
```

in a code cell and hit [shift]+[enter]. A help window pops up at the bottom of the screen. Click on the x in the upper right of the help window to make it disappear.

In the example that follows, the print function is used to display the value of variable b on the screen

```
b = 6
print(b)
```

6

In the previous code cell, simply typing b on the last line also works, as explained previously. When using the print command, Out[n]: (n being a number) does not appear in front of what is being printed. The print function can take multiple arguments

```
print(a5.real, a5.imag)
```

2.0 3.0

String arguments may be mixed with variables

```
print('The value of a is', a)
```

The value of a is 2

It is better (and nicer) to combine text and variables using f-strings. The string must be preceded by the letter f, and the variables that are printed are enclosed by braces {}

```
print(f'The value of a is {a}')
```

The value of a is 2

A list is a Python data structure that contains a collection of data that can be of any type. It is defined by typing the elements of the list separated by commas and enclosed by square brackets. The following list contains a string, an integer, and a float.

```
alist = ['Darcy', 0, 2.3]
print(alist)
```

['Darcy', 0, 2.3]

An individual item in a list can be accessed by specifying the index in the list between square brackets. Selecting elements like this is called indexing or slicing, which is discussed in more detail in the section on arrays. Remember that Python counts using a zero-based system, so the first element in the list has the index number 0. A range of items can be selected by specifying the first index and the last index separated by a colon

```
print(alist[0])
print(alist[0:2]) # Returns items 0 and 1, but not item 2
```

```
Darcy
['Darcy', 0]
```

The second print statement above only returns items 0 and 1, but not item 2. This is because of the way the counting works in Python: It starts at zero and then stops before the last number is reached. The first item in alist can be changed as follows

```
alist[0] = 'Dupuit'
print(alist)
```

```
['Dupuit', 0, 2.3]
```

Items can also be appended to a list. For example, the number 12 can be added to alist

```
alist.append(12)
print(alist)
```

```
['Dupuit', 0, 2.3, 12]
```

The + operator can be used to combine two lists

```
a = (1, 2, 3)
b = (5, 10, 15)
print(a + b)
```

```
(1, 2, 3, 5, 10, 15)
```

Note that the use of the + sign in the last code cell does not imply an arithmetic operation. The numbers in the lists are not added. If the intent is to add the numbers, the lists must first be converted to arrays, which are discussed later on.

A list is one of several basic sequence types in Python. Another one is the tuple, which, like a list, is a sequence of items, but it is defined by enclosing the elements between parentheses rather than square brackets. The difference between a list and a tuple is that elements in a tuple cannot be changed after the tuple has been defined.

A.2   Loops and if statements

In Python, a for loop is a set of programming commands that is executed several times. A for loop is explained here by looping over all the items in a list.

There are two syntax rules for defining a for loop. First, a colon (:) must always terminate the line that starts with for, and second, the lines of code that must be executed during each step of the loop must all be indented. This is illustrated in the following example, in which the variable item takes on the value of each element in the list alist during the for loop. The second, indented line tells Python to print the value as well as the variable type of item for each step of the loop

```
for item in alist:
    print(item, 'has type', type(item))
```

```
Dupuit has type <class 'str'>
0 has type <class 'int'>
```

```
2.3 has type <class 'float'>
12 has type <class 'int'>
```

There is no explicit command that marks the end of the code that must be executed during a loop. Instead, Python's interpreter infers this based on the indentation. Proper indentation is thus essential as it controls the way the code behaves. This helps tremendously in creating neat, readable scripts. The length of the indentation is arbitrary, as long as it remains the same; the default is four spaces, so use that.

Looping over the elements of a list is a very pythonic way of doing things. Programmers using other languages may be used to loops that get executed a certain number of times. This is possible too and is typically done using the range command. In the following example, range(4) creates 4 numbers, from 0 up to (but not including) 4

```
for i in range(4):
    print(f'Step {i}')
```

```
Step 0
Step 1
Step 2
Step 3
```

The range(n) command creates a sequence with n numbers to iterate over, starting at 0. range can optionally be defined by specifying a starting value (the default is zero) and a step (the default step is 1). So a range starting at -3 and stopping before the counter reaches +3 with steps of 2 is defined as

```
for k in range(-3, 3, 2):
    print(k)
```

```
-3
-1
1
```

Python also provides the enumerate function that generates an index counter as it loops through a list. This can be useful for example when the corresponding items in another list have to be modified during the loop. In the following example, the strings in the list names are combined to full names using the strings in the list last_names. By including enumerate() in the line that starts the for loop, a counter i is created so that each time the for loop steps to the next element in the list last_names, i is incremented by 1. Notice the syntax: The counter i is a user-defined variable name that precedes the name that is used within the for loop for the active element in the list (last_name in this case).

```
names = ["Henry", "Willem", "Alexander"]
last_names = ["Darcy", "Badon Ghyben", "Herzberg"]
print("First names", names)
for i, last_name in enumerate(last_names):
    names[i] = names[i] + " " + last_name
print("Full names", names)
```

```
First names ['Henry', 'Willem', 'Alexander']
Full names ['Henry Darcy', 'Willem Badon Ghyben', 'Alexander Herzberg']
```

Conditional statements (also called if statements) can be used to control the flow of the program depending on whether or not certain conditions are met. Just like with the for loop, a colon must be typed at the end of the line starting with if, and the lines that must be executed if the evaluated condition is true must be indented. There is no specific command to end the if statement: As with for loops it is inferred from the indentation.

In the next code cell, it is checked if the list alist actually contains elements by calling the len function, which returns the number of items (zero for an empty list).

```python
alist = [] # Defines an empty list
if (len(alist) == 0):
    print ("alist contains no elements")
```

```
alist contains no elements
```

Note that a double equal sign is used for the comparison (since a single equal sign is reserved for value assignment). Other comparison operators in Python include the following (note that the comparisons return a Boolean variable type, which can be printed to the screen).

```python
a = 4
print(a <= 4) # a is smaller than or equal to 4
print(a == 4) # a is equal to 4.
print(a >= 4) # a is larger than or equal to 4
print(a != 4) # a is not equal to 4
```

```
True
True
True
False
```

Conditions can be compounded with the and or or statements, with parentheses for readability

```python
a = 7
if (a > 3) and (a < 8):
    print('The value of a is', a)
```

```
The value of a is 7
```

In the following code example, the if statement is placed inside a for loop to determine what message is printed to the screen. The message varies according to the value of the loop variable i. The if statement now also includes the elif and else commands (note how the indentation works)

```python
for i in range(-1, 3):
    if i < 0:
        print(i, 'is negative')
    elif i == 0:
        print(i, 'is zero')
    else:
        print(i, 'is positive')
```

```
-1 is negative
0 is zero
1 is positive
2 is positive
```

A.3   The numpy package and arrays

The core of Python includes the basic features of the language. All other features are included in separate packages (called libraries or toolboxes in other programming languages). The package for creating and manipulating arrays is called numpy. The numpy package also includes a set of basic mathematical functions (cos, exp, log, sum, etc.). Extensive resources on numpy's capabilities are available on the Internet, including the numpy web site. This section focuses on how to use numpy arrays of numbers.

Before a package can be used, it needs to be imported. The following line imports the numpy package and renames it to np (the default short name used in the scientific Python community).

```
import numpy as np
```

All functions of the numpy package can now be called using np.function_name(), for example

```
np.sqrt(10)
```

3.1622776601683795

Mathematical numpy functions work on both real and complex variables. Note, however, that the square root of a negative number returns a nan (not a number) and a warning is thrown

```
np.sqrt(-1)
```

```
<ipython-input-48-597592b72a04>:1: RuntimeWarning: invalid value encountered in
sqrt
  np.sqrt(-1)
```

nan

A complex variable is returned when a complex variable is passed to the function (even when the imaginary part equals zero)

```
np.sqrt(-1 + 0j)
```

1j

There are multiple ways to create an array. For example, a list can be converted to a numpy array with the np.array() command

```
a = np.array([1, 2, 3, 4])
print(a, 'has type', type(a))
```

[1 2 3 4] has type <class 'numpy.ndarray'>

The type function shows that a is an ndarray, which stands for n-dimensional array. All elements within such an array have the same variable type (contrary to a list, which can contain items of different types).

Functions that work for individual numbers can also be used on arrays, for example

```
print('sqrt(a) gives:', np.sqrt(a))
```

sqrt(a) gives: [1.   1.41421356 1.73205081 2.   ]

Arithmetic operators can also be used with arrays. The following code defines an array b and multiplies the elements of the arrays a and b element by element

```
b = np.array([2, 2, 3, 3])
print('a * b gives:', a * b)
```

a * b gives: [2 4 9 12]

Readers working with matrices may want to compute the dot product of two vectors (arrays), rather than multiplying two vectors term by term. The dot product of arrays a and b is obtained by using the @ operator

```
a @ b # Or, alternatively: np.dot(a,b)
```

27

In addition to converting a list to an array using the np.array() function, arrays can be generated with the np.arange function (similar to the range function used earlier, but now resulting in an array). The arange(start, end_before, step) function creates an array starting at start, taking steps equal to step, and stopping before it reaches the specified end_before value. If only a single value is passed to the function, it is interpreted to be the end_before value (i.e., the default values for start and step are 0 and 1, respectively). Using a negative step is possible as long as the start argument is larger than the end_before argument. Note that the default type of np.arange is integers unless one of the arguments is a float.

```
print(np.arange(3)) # so start=0 and step=1
print(np.arange(2, 6)) # so step=1
print(np.arange(2, 10, 2))
print(np.arange(10, 0, -2)) # steps can be negative
print(np.arange(2, 10, 2.0)) # now returns array of floats
```

```
[0 1 2]
[2 3 4 5]
[2 4 6 8]
[10  8  6  4  2]
[2. 4. 6. 8.]
```

Instead of specifying a step, it is also possible to create an array by specifying the number of elements using the np.linspace(begin, end, num) function. The following function call returns seven points with values of equal increment from 1 to 4. Note that, somewhat confusingly, and unlike range and np.arange, the end value in linspace is included as the last element of the array. The default return value of np.linspace is an array of floats.

```
np.linspace(1, 4, 7)
```

```
array([1. , 1.5, 2. , 2.5, 3. , 3.5, 4. ])
```

Arrays may have multiple dimensions. A two-dimensional array can be thought of as a matrix with rows and columns. It can be created by entering a list of rows, where each row is again a list that contains the values for that row. Care must be taken that each of the lists has the same number of elements (equal to the number of columns). When the array is printed to the screen, each row is on a separate line and the columns are separated by a space.

```
A = np.array([[0, 1, 2, 3], [4, 5, 6, 7], [8, 9, 10, 11]])
print(A)
```

```
[[ 0  1  2  3]
 [ 4  5  6  7]
 [ 8  9 10 11]]
```

Arrays filled with zeros or ones may be created with the functions np.zeros and np.ones, respectively. Each of these functions requires the specification of a shape (e.g., the length of the array, or a tuple with the number of rows and columns for a two-dimensional array). The default return type is float.

```
print(np.zeros(7))
print(np.ones((2, 3))) # 2 rows, 3 columns
```

```
[0. 0. 0. 0. 0. 0. 0.]
[[1. 1. 1.]
 [1. 1. 1.]]
```

Just like for a list, the individual elements of an array can be accessed by indexing using square brackets (remember that the first index has number 0). The syntax for a one-dimensional array x is x[start:end_before] or x[start:end_before:step]. If the start isn't specified, 0 will be used. If the step isn't specified, 1 will be used.

When programming, the index of the end is generally not known as the array size can vary. To find out how long the array is and access the last value of array x, one can type x[len(x) - 1], but that is a bit inconvenient. Luckily, there is a shortcut: x[-1] is the same as x[len(x) - 1] and represents the last value in the array. Typing x[-2] returns the second to last value in the array, etc. The following code examples demonstrate various slicing operations.

```
x = np.arange(20, 30)
print(x)
print(x[0])
print(x[0:5]) # same as x[:5]
print(x[-5:]) # returns the last 5 values
print(x[3:7])
print(x[2:9:2]) # step is 2
print(x[-1:4:-2]) # starts at back, stops before reaching index 4 with step -2
print(x[[0, 3, 7]]) # passes indices as a list
```

```
[20 21 22 23 24 25 26 27 28 29]
20
[20 21 22 23 24]
[25 26 27 28 29]
[23 24 25 26]
[22 24 26 28]
[29 27 25]
[20 23 27]
```

The contents of an array can be modified by assigning values to an array slice. In the following example, the first 5 elements of array x are replaced by 40, after which the last value is changed to 40 in a second command.

```
x = 20 * np.ones(10)
print(x)
x[0:5] = 40
print(x)
x[-1] = 40 # change last value
print(x)
```

```
[20. 20. 20. 20. 20. 20. 20. 20. 20. 20.]
[40. 40. 40. 40. 40. 20. 20. 20. 20. 20.]
[40. 40. 40. 40. 40. 20. 20. 20. 20. 40.]
```

Parts or values of an array may be selected based on conditions. This is called advanced slicing. For example,

```
a = np.arange(5)
print(a)
print(a < 3)  # a < 3 returns a Boolean array
```

```
[0 1 2 3 4]
[ True   True   True False False]
```

Because < is a conditional operator, the statement a < 3 returns a Boolean array. Instead of printing the outcome to the screen, as in the above example, a more useful application is to store it as a variable. The variable can then be used to slice the array.

```
a_less_than_3 = a < 3
print('values in array a less than 3:', a[a_less_than_3])
```

```
values in array a less than 3: [0 1 2]
```

The ~ operator swaps the True and False values

```
print('values in array not less than 3:', a[~a_less_than_3])
```

```
values in array not less than 3: [3 4]
```

For more compact and readable code, the conditional statement can be specified inside the square brackets. It can even be used to assign a different value to selected elements of an array

```
a[a < 3] = 10   # Replace all values less than 3 by 10
print('new a:', a)
```

```
new a: [10 10 10  3   4]
```

Conditions for indices can be compounded with either & (and) or | (or), when the conditions are put between parentheses.

```
a = np.arange(10)
print('array:', a)
print('less than 3 or larger than 7:', a[(a < 3) | (a > 7)])
print('larger than 3 and less than 7:', a[(a > 3) & (a < 7)])
```

```
array: [0 1 2 3 4 5 6 7 8 9]
less than 3 or larger than 7: [0 1 2 8 9]
larger than 3 and less than 7: [4 5 6]
```

Arrays can be stacked next to each other (horizontally) using the np.hstack command and above each other (vertically) using the np.vstack function

```
a = np.arange(3)
print(np.hstack((a, a)))
print(np.vstack((a, a)))
```

```
[0 1 2 0 1 2]
[[0 1 2]
 [0 1 2]]
```

Arrays can be combined to form larger arrays in this way, but it is not very efficient computationally, so programs may run slowly when large arrays are stacked or arrays are stacked multiple times, for example in a loop. It is better to create an array of the correct size first and then fill it with a for loop (this may be difficult when the size of the array is not known upfront). Note the use of enumerate in the code cell below.

```
x = np.linspace(0, np.pi, 4)
y = np.empty(len(x))
for i, xx in enumerate(x):
    y[i] = np.sin(xx)
print(y)
```

```
[0.00000000e+00 8.66025404e-01 8.66025404e-01 1.22464680e-16]
```

The code cell above is a demonstration of creating an empty array and filling it with a `for` loop. For this specific example, `y = sin(x)` is, of course, much easier and faster.

A.4   The `matplotlib` package for visualization

The main package for plotting in Python is called `matplotlib`. It is a comprehensive plotting package that is highly versatile and can be used to make fantastic plots and animations. The gallery on the `matplotlib` web page is a great resource to discover many of the `matplotlib` capabilities.

Most of `matplotlib`'s plotting capabilities are available in the subpackage `pyplot`, which is commonly renamed as `plt` in the Python community. All figures can be included inline in a notebook (rather than in a separate window) with the IPython magic command `%matplotlib inline`. IPython magic commands are preceded with a `%`. They are not standard Python commands and only work within IPython environments. They are intended to provide functionality that makes programming a little easier.

```
%matplotlib inline
import matplotlib.pyplot as plt
```

The following two lines of code compute $\sin(x)$ for 100 values of x between 0 and π, and then plot $\sin(x)$ vs. x.

```
x = np.linspace(0, np.pi, 100)
plt.plot(x, np.sin(x));
```

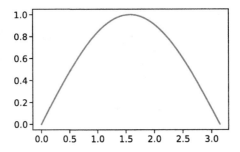

The plot function has many, many options, as may be seen from the documentation. A more complex graph with multiple lines, labels along the axes, and a legend is shown below. `matplotlib` uses a different color for each line that is plotted. The default colors are called `C0`, `C1`, `C2`, etc. Alternatively, a different color can be specified.

```
x = np.linspace(0, 2 * np.pi, 200)
y1 = np.sin(x)
y2 = np.cos(x)
plt.plot(x, y1, '--', label='sin(x)')         # plotted with default color C0
plt.plot(x, y2, color='purple', lw=2, label='cos(x)')   # linewidth (lw) is 2
```

```
plt.xlabel('x-axis')
plt.ylabel('y-axis')
plt.legend(); # selects best spot for legend unless loc is specified
```

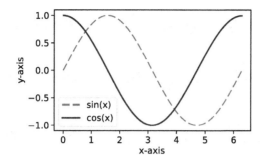

Most `matplotlib` functions return a handle to the plot item that is created. For example, the `plt.plot` function returns a list with a handle to the line that is plotted

```
a = plt.plot([1, 2, 3], label='hello')
print(a)
```

```
[<matplotlib.lines.Line2D object at 0x116b51c40>]
```

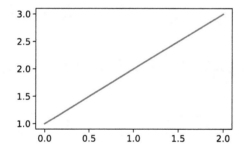

The cryptic statement such as `[<matplotlib.lines.Line2D at 0x10a9063d0>]` is a list with a handle to the memory location of the line that was plotted. For example, the command `plt.xlabel` adds a text object and returns a handle to this object. The properties of the item that the handle refers to can be modified with the `plt.setp` command (although this is not done frequently in this book). In the following example, the `setp` function is used to increase the line width and color of the plotted line.

```
a = plt.plot([1, 2, 3])
plt.setp(a, color='hotpink', lw=4);
```

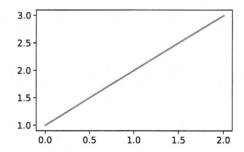

Figures can be saved to a file using `plt.savefig('/mydirectory/myfigure1.png')` where `mydirectory` is the directory name. Forward slashes must be used to separate directory names. The file type is inferred from the file extension (png, eps, and svg).

Two-dimensional data can be visualized with a contour plot. For example, consider the following function for the head h as a function of the horizontal coordinates x and y

$$h = \frac{2}{\pi} \ln\left(x^2 + y^2\right) - 0.01x - 5$$

A contour plot of the head is created in a window with lower-left corner $(x, y) = (-200, -200)$ and upper-right corner $(x, y) = (400, 200)$. A grid of (x, y) values is defined first using the `np.meshgrid` function, which creates the two-dimensional arrays x and y. The head is computed at each (x, y) location. Before contouring, a figure is created with `plt.subplot` and using the keyword `aspect=1` to make sure that the horizontal and vertical axes have the same scale (so that a circle looks like a circle rather than an ellipse). The head values are contoured with the `plt.contour` function; 20 head values are contoured.

```
x, y = np.meshgrid(np.linspace(-200, 400, 100), np.linspace(-200, 200, 100))
h = 2 / np.pi * np.log(x ** 2 + y ** 2) - 0.01 * x - 5
plt.subplot(111, aspect=1)
plt.contour(x, y, h, 20);    # draw 20 contour lines
```

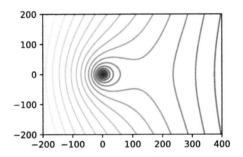

The head values that are contoured can be specified with a list or array. First, the minimum and maximum values of the computed head are found with the `np.min` and `np.max` functions.

```
print('min:', np.min(h))
print('max:', np.max(h))
```

```
min: -4.1046611143428455
max: 4.18729839195133
```

The head is contoured for values from -3 m up to 6 m with steps of 0.5 m. The handle that is returned by the `plt.contour` function is stored and passed to the `plt.clabel` function to plot labels along the contours. The format of the labels is specified as `fmt='%.1f'`, which stands for floating point numbers with 1 decimal place.

```
plt.subplot(111, aspect=1)
cs = plt.contour(x, y, h, np.arange(-3, 6, 0.5), colors='C0')
plt.clabel(cs, fmt='%.1f');
```

The color of all contour lines is specified as `colors='C0'`, and the contour lines for negative values are, by default, dashed. This latter behavior can be overridden by specifying the keyword argument `linestyles='-'` of `plt.contour` or by changing the value of the default `matplotlib` parameter `contour.negative_linestyle` (see the `matplotlib` documentation).

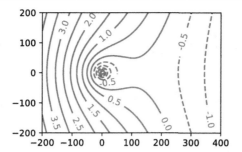

A.5   Functions

Functions are defined in Python with the def command. Functions may be stored in files, or they can be typed in code cells in the notebook or at the Python prompt in a console. The following function is implemented in the Python function func_one below.

$$f(x) = \cos(x) \qquad x < 0$$
$$f(x) = \exp(-x) \qquad x \geq 0$$

```
def func_one(x):
    if x < 0:
        rv = np.cos(x)
    else:
        rv = np.exp(-x)
    return rv
```

The name of the function is func_one (it is proper Python coding style to use lowercase letters for functions and to separate words by underscores). The function returns the value of the variable rv that is computed inside the function. Typing func_one(followed by [Shift][Tab] in a code cell displays the input arguments of the function (and any documentation, if provided). Calling the function returns the desired values

```
print(func_one(1))
print(func_one(-1))
```

```
0.36787944117144233
0.5403023058681398
```

The function func_one cannot be called with an array because of the if-else construct, which only works for single numbers. There are two approaches to make func_one work on an input array. The first and easiest way is to create a new function that vectorizes the func_one function. The new function takes an array as its argument and simply calls the func_one function for every value of the array and then returns an array. A function may be vectorized with the np.vectorize function, which takes as input argument a function to be vectorized and returns a function.

```
func_two = np.vectorize(func_one)
func_two(np.arange(-2, 3))  # func_two can be called with an array
```

```
array([-0.41614684,  0.54030231,  1.        ,  0.36787944,  0.13533528])
```

The use of np.vectorize is easy, but not very fast. Behind the scenes, the func_two function simply creates a loop and loops over all the values in the array. Alternatively, the func_one function can be rewritten to accept an array as input argument. The `np.atleast_1d` function can be used to make sure that the input argument x is at least a one-dimensional array (so a scalar is converted to an array). The `np.empty(x.shape)` function is used to create an empty array of the same shape as x. Conditional statements are used to fill the array, which is returned at the end of the function.

```
def func_one_array(x):    # fancy function that works on arrays
    x = np.atleast_1d(x)
    f = np.empty(x.shape) # return value is now called f
    f[x < 0] = np.cos(x[x < 0])
    f[x >= 0] = np.exp(-x[x >= 0])
    return f
```

The `func_one_array` function is used, and a graph is created below. Note that the limits of the axes are set and fancy tick marks are added using list comprehension and a LaTex-style command to plot the Greek letter π.

```
x = np.linspace(-2 * np.pi, 2 * np.pi, 201)    # x values
y = func_one_array(x) # call the function to compute the y values
plt.plot(x, y)
plt.xlim(-2 * np.pi, 2 * np.pi)    # set the limits along the x-axis
plt.xticks(np.arange(-2 * np.pi, 3 * np.pi, np.pi),
           ['$-2\pi$', '$-\pi$', '0', '$\pi$', '$2\pi$']);
```

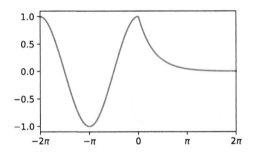

Documentation of the function can be provided inside the function using triple quotes. This is, of course, proper programming style, but is not used in this book to save space.

```
def func_one(x):
    '''Documentation of the function goes here
    Returns:
    cos(x) for x < 0
    exp(-x) for x >=0'''
    if x < 0:
        rv = np.cos(x)
    else:
        rv = np.exp(-x)
    return rv
```

Typing `func_one(` followed by [Shift][Tab] now shows the documentation.

A value that is passed to a function, like x in the previous examples, is called an argument. Functions do not always have an argument. Function arguments can be regular arguments and keyword arguments (or `kwargs`). Regular arguments always need to be passed and must be in the proper order. Keyword arguments are arguments that are optional for the user to pass, and the order in which they are passed does not matter. If a keyword argument is not passed, its value takes the default value that is specified in the function definition. When a function has both regular arguments and keyword arguments, the keyword arguments must always come after the regular arguments. These rules are illustrated in the following example where the function test has one regular and two keyword arguments.

```
def test(x, y=7, z=100):
    print(x, y, z)

test(3)
test(3, y=8) # or test(3, 8), but better to specify keyword
test(3, z=-1)
test(3, z=-1, y=8) # keyword arguments may be in any order
test(x=3) # regular arguments may also be called with their name
```

```
3 7 100
3 8 100
3 7 -1
3 8 -1
3 7 100
```

If a function takes no arguments, its definition must still include the parentheses. Similarly, calling a function also requires adding the parentheses.

```
def test():
    print('Hello')
test()
```

```
Hello
```

If a variable is used inside a function, but the variable is not explicitly passed to the function with a regular argument or a keyword argument, Python will search for this variable in the memory outside the function. This is generally considered sloppy programming, but it is sometimes done in short scripts, also in this book. So the following function works, but is not optimal.

```
velo = 0.05 # defined outside the function

def printflow(x):
    print(f'the velocity at x={x} is {velo} m/d')

printflow(100)
```

```
the velocity at x=100 is 0.05 m/d
```

It is better to pass all variables to the function

```
def printflow(x, velo):
    print(f'the velocity at x={x} is {velo} m/d')

printflow(x=100, velo=0.05)
```

```
the velocity at x=100 is 0.05 m/d
```

It can be useful to define a function in such a way that any keyword argument may be passed to it, even if it is not known a priori or explicitly defined. This is useful if the function calls another function. The code cell below demonstrates how this works by defining a function kwargs_demo that creates a figure with a fixed layout, but with user-defined options for the way the data are plotted. By including **kwargs in the list of arguments for kwargs_demo, and passing them on to the `plot` function, all of `matplotlib`'s `plot` keyword arguments become available for use with the `kwargs_demo` function.

```
def kwargs_demo(x, y, **kwargs):
    plt.figure()
    plt.plot(x, y, **kwargs)
    plt.grid(True)
    plt.xlabel("x")
    plt.ylabel("y")

x = np.linspace(0, 4 * np.pi)
y = np.cos(x)
kwargs_demo(x, y, color='C2', ls='--', lw=4);
```

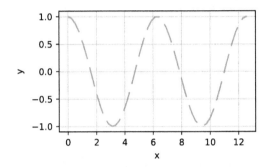

Python code or functions can be defined in a separate file with extension .py. Such a file can contain one or many function definitions. Storing functions this way avoids the need of having to define them repeatedly, and it allows functions to be shared and distributed easily. A set of functions that perform a similar task is called a *package*. Packages can consist of many different files and can have a complex structure, which is not discussed here.

There are many different ways to import packages. Using the numpy package as an example, the most basic syntax is `import numpy` after which any function in numpy can be called as `numpy.function_name()`. Alternatively, all of numpy's functionality may be imported with `from numpy import *` This allows numpy's functions to be called without prefixing numpy to the function name. However, importing libraries this way is generally discouraged as any relation between the functions and packages is lost. A better way is to use `import numpy as np` (as used before and as is standard practice in the Python community). This makes it possible to write compact code while maintaining the relationship between the package and everything defined in it. This is important for programs that rely on multiple packages, as similar names may exist across packages. By prefixing the package name to a function name, the programmer has certainty that the intended function indeed gets called.

Packages can also have subpackages. The `pyplot` subpackage of `matplotlib` was already imported earlier using `import matplotlib.pyplot as plt`. Another example of a subpackage is the `linalg` package of `numpy`, which is a collection of functions that deal with linear algebra. If the numpy package is imported with `import numpy as np`, the functions in the `linalg` subpackage can be called `np.linalg.function_name()`.

A.6   The `scipy` package for scientific computing

The three main Python packages for exploratory computing are numpy, `matplotlib`, and scipy. The numpy and `matplotlib` packages were already discussed in previous examples.

Here, some of the functionalities of `scipy` are demonstrated, including root-finding, minimization, special functions, and numerical integration.

The function `fsolve` from the `scipy.optimize` package can be used to find the root or roots of a function. In the code cell below, the root of the function $f(x) = e^{x/5} - 2$ is found numerically using the `fsolve` function (it can, for this case, also be computed analytically, of course). A starting value for the numerical search must be supplied. The closer the starting value to the root, the more likely (and the quicker) `fsolve` finds the root.

```
def ffunc(x):
    return np.exp(x / 5) - 2

from scipy.optimize import fsolve
xroot = fsolve(ffunc, x0=0) # start searching at x0=0
print(f'root={xroot}, function value={ffunc(xroot)}')
```

```
root=[3.4657359], function value=[2.7533531e-14]
```

Next, three points are found where $f(x)$ intersects the function $g(x) = \sin(x)$. The function `fdiff` computes the difference between $f(x)$ and $g(x)$, and `fsolve` is used to find three roots of `fdiff` for three different starting locations; that is, `fsolve` finds three points where the two functions intersect.

```
def gfunc(x):
    return 3 * np.sin(x)

def fdiff(x):
    return ffunc(x) - gfunc(x)

x_intersect = fsolve(fdiff, np.array([3, 6, 9])) # 3 starting points
x = np.linspace(0, 3 * np.pi)
plt.plot(x, ffunc(x), label="f(x)")
plt.plot(x, gfunc(x), label="g(x)")
plt.plot(x_intersect, ffunc(x_intersect), 'k.')
plt.legend();
```

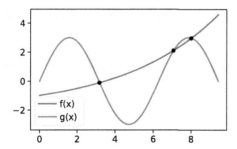

The minimum of a function may be found with the `fmin` function of `scipy.optimize`. The code cell below finds the x-location where the difference between the functions $g(x)$ and $h(x) = e^{x/5} + 2$ is the smallest, starting the search at $x = 1.5$.

```
def hfunc(x):
    return np.exp(x/5) + 2

def fdiff(x):
    return hfunc(x) - gfunc(x)
```

```
from scipy.optimize import fmin
xmin = fmin(fdiff, 1.5)

plt.plot(x, hfunc(x), label="h(x)")
plt.plot(x, gfunc(x), label="g(x)")
plt.plot(xmin, hfunc(xmin), 'C0.')
plt.plot(xmin, gfunc(xmin), 'C1.')
plt.legend();
```

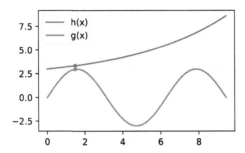

Two other subpackages of the scipy package are scipy.special for special functions and scipy.integrate for numerical integration. As an example, consider the exponential integral defined as

$$E_1(u) = \int_u^\infty \frac{e^{-s}}{s} ds \qquad (A.1)$$

E_1 is available as exp1 from the scipy.special package, but can also be evaluated using numerical integration. The quad function of the scipy.integrate package is used and takes as input the function to integrate and the lower and upper limits of the integration; use np.inf for ∞. Note that quad returns two values: the outcome of the numerical integration and an error estimate. The results of the numerical integration and the special function are compared below for $E_1(1)$.

```
from scipy.integrate import quad
from scipy.special import exp1

def integrand(s):
    return np.exp(-s) / s

E1num = quad(integrand, 1, np.inf)[0]  # numerical integration

print('E1(1) from numerical integration:', E1num)
print('E1(1) from special function      :', exp1(1))
```

```
E1(1) from numerical integration: 0.21938393439551238
E1(1) from special function      : 0.21938393439552062
```

Numerical answers to selected problems

Chapter 0
0.1 $h = 18.06$ m
0.2 $k = 11.29$ m/d, $k = 16.93$ m/d
0.3 $v = 0.13$ m/d $= 4.0$ m/month

Chapter 1
1.1 $x_d = 300$ m
1.2 $h_0 = 9$ m
1.3 $N = -0.0008$ m/d
1.6 $t_{tr} = 2079$ d
1.7 $t_{tr} = 2079$ d
1.9 $h_L = 4.64$ m
1.10 $t = 20625$ d at $x = L_0$, $t = 41250$ d at $x = L$ for $H_0 = H_1 = 10$ m
$t = 30937$ d at $x = L$ for $H_1 = 5$ m.
1.11 $h_0 = 7.75$ m

Chapter 2
2.1 $c = 52.0$ d
2.4 $U_L = 0.0574$ m^2/d,
$U_R = -0.0936$ m^2/d
2.5 $T_{eff} = 127.3$ m^2/d

Chapter 3
3.1 $H = 11.227$ m
3.5 $h = 6.550$ for $H = 10$ m
$h = 6.446$ m for $H = 20$ m

Chapter 4
4.1 $U = 0.5$ m^2/d, $x_{toe} = -100$ m
4.2 $U = 0.06$ m^2/d, $x_{toe} = -800$ m
4.3 $x_{toe} = -125$ m
4.4 $x_{toe} = -886$ m
4.5 $x_{toe} = 500 \pm 212.13$ m before sea level rise. $x_{toe} = 500 \pm 154.88$ m after sea level rise
4.6 $N = 0.00164$ m/d
4.7 $h_0 = 1.11$ m with interface
$h_0 = 0.62$ m without interface
4.8 Toe moves inland 25.05 m after sea level rise

4.9 $h = 0.11$ m, $z_i = -4.22$ m at shoreline
$h = 0.404$ m, $z_i = -16.18$ m at center of island

Chapter 5
5.1 $T = 1$ m^2/d
5.2 $x = 1000$ m
5.5 $V = 31.9$ m^3/m after 10 days
$V = 8.6$ m^3/m after 40 days
5.7 $A = 218.85$ d, $a = 42.56$ d
5.8 $h = 0.618$ m at $t = 0.01$ d, $h = 1.356$ m at $t = 0.1$ d, $h = 1.458$ m at $t = 1$ d

Chapter 6
6.1 $r_d = 141.4$ m
6.2 $Q = 62.83$ m^3/d
6.3 $Q = 251.2$ m^3/d
6.4 $Q = 686$ m^3/d
6.5 $h_w = -3.68$ m
6.6 $\Psi = [-25, -50, 50, 25]$ m^3/d
6.7 $h = -1.04$ m at extraction well
$h = 1.18$ m at injection well.
6.8 $Q = 432.6$ m^3/d
6.9 $h_w = 7.12$ m
6.11 $h_0 = -4.81$ m, $h_1 = -3.15$ m for $c = 200$ d; $h_0 = -5.18$ m, $h_1 = -3.41$ m for $c = 1000$ d
6.12 $h_0 = 4.93$ m, $h_1 = 1.60$ m

Chapter 7
7.1 $Q_x = 0.4$ m^2/d, $Q_y = 0.2$ m^2/d
7.2 $Q = 26.7$ m^3/d
7.3 $x_s = \pm 102.3$ m
7.4 $x_{s_1} = -24.1$ m, $x_{s_2} = 103.7$ m
7.7 $Q = 125.7$ m^3/d
7.8 $h = 10.4$ m
7.9 $h = -1.73$ m

Chapter 8

8.2 $h = 18.8$ m, $Q = 296$ m^3/d
8.4 $Q = -1898$ m^3/d
8.5 $Q = 387$ m^3/d

Chapter 9

9.1 $t = 9$ d
9.2 $r = 9.27$ m
9.6 $T = 256$ m^2/d, $S = 0.057$ without river; $T = 197$ m^2/d, $S = 0.107$ with river
9.8   Theis: $T = 4484$ m^2/d, $S = 0.0018$, $RMSE = 0.12$ m

Hantush: $T = 3135$ m^2/d, $S = 0.0044$, $c = 25.2$ d, $RMSE = 0.02$ m
9.9 $T = 560$ m^2/d, $S = 4.16 \cdot 10^{-5}$

Chapter 10

10.3 $\Delta h = 0.0775$ m
10.4   Dupuit: $\Delta h = 0.0750$ m, extra head drop: $\Delta h = 0.0025$ m
10.6   No regional flow cell for $H = 70$ m
10.7   Factor based on head at the top of aquifer at $x = -100$ m: 41.6. Factor based on head at the top of aquifer at $x = -50$ m: 43.1
10.8   Factor based on head at the top of aquifer at $x = 0$ m: 38.5

Bibliography

Abramowitz, M. and Stegun, I. (1965), *Handbook of mathematical functions with formulas, graphs, and mathematical tables*, 9th edn, Dover.

Anderson, M. P., Woessner, W. W. and Hunt, R. J. (2015), *Applied groundwater modeling: simulation of flow and advective transport*, Academic press.

Bakker, M. (2006), 'Analytic solutions for interface flow in combined confined and semi-confined, coastal aquifers', *Advances in Water Resources* **29**(3), 417–425.

Bakker, M. (2013*a*), Analytic modeling of transient multilayer flow, *in* 'Advances in hydrogeology', Springer, pp. 95–114.

Bakker, M. (2013*b*), 'Semi-analytic modeling of transient multi-layer flow with ttim', *Hydrogeology Journal* **21**(4), 935–943.

Bakker, M. (2021), *TimML, A multi-layer, analytic element model.* https://github.com/mbakker7/timml.

Bakker, M. and Anderson, E. I. (2003), 'Steady flow to a well near a stream with a leaky bed', *Groundwater* **41**(6), 833–840.

Bakker, M. and Kelson, V. A. (2009), 'Writing analytic element programs in Python', *Groundwater* **47**(6), 828–834.

Bakker, M. and Schaars, F. (2019), 'Solving groundwater flow problems with time series analysis: You may not even need another model', *Groundwater* **57**(6), 826–833.

Bakker, M. and Strack, O. D. (2003), 'Analytic elements for multiaquifer flow', *Journal of Hydrology* **271**(1-4), 119–129.

Bear, J. and Jacobs, M. (1965), 'On the movement of water bodies injected into aquifers', *Journal of Hydrology* **3**(1), 37–57.

Bear, J. and Verruijt, A. (1987), *Modeling Groundwater Flow and Pollution*, D. Reidel Publishing Company.

Bruggeman, G. (1999), *Analytical Solutions of Geohydrological Problems*, Vol. 46 of *Developments in Water Science*, Elsevier.

Carr, P. A. and Van Der Kamp, G. S. (1969), 'Determining aquifer characteristics by the tidal method', *Water Resources Research* **5**(5), 1023–1031.

Collenteur, R. A., Bakker, M., Caljé, R., Klop, S. A. and Schaars, F. (2019), 'Pastas: Open source software for the analysis of groundwater time series', *Groundwater* **57**(6), 877–885.

Craig, J. R. (2008), 'Analytical solutions for 2D topography-driven flow in stratified and syncline aquifers', *Advances in water resources* **31**(8), 1066–1073.

Darcy, H. (1856), *Les Fontaines Publiques de la Ville de Dijon*, Dalmont, Paris.

De Glee, G. (1930), *Over grondwaterstroomingen bij waterontrekking door middel van putten*, Delft University. PhD thesis.

Dupuit, J. (1863), *Études théoriques et pratiques sur le mouvement des eaux dans les canaux découverts et à travers les terrains perméabls: avec des considérations relatives au régime des grandes eaux, au débouché à leur donner, et à la marche des alluvions dans les rivières à fond mobile*, Dunod, Paris.

Edelman, F. (1947), *Over de berekening van grond water stromingen*, Delft University. PhD thesis.

Ferris, J. G. (1952), *Cyclic fluctuations of water level as a basis for determining aquifer transmissibility*, US Department of the Interior, Geological Survey, Water Resources Division, Ground Water Branch.

Fitts, C. (2013), *Groundwater science*, second edn, Academic Press.

Fitts, C. R. (2010), 'Modeling aquifer systems with analytic elements and subdomains', *Water Resources Research* **46**(7).

Fitts, C. R. (2021), *AnAqSim (analytic aquifer simulator)*. `https://www.fittsgeosolutions.com`.

Forchheimer, P. (1886), 'Über die ergiebigkeit von brunnen-anlagen und sickerschlitzen', *Z. Architekt. Ing.-Ver. Hannover* **32**, 539–563.

Forchheimer, P. (1919), 'Zur theorie der grundwasserstroemung', *Sitzungsberichten der Akademie der Wissenschaften in Wien, Mathem.-natunw. Klasse, Abteilung IIa* **128**, 7–14.

Glover, R. E. (1959), 'The pattern of fresh-water flow in a coastal aquifer', *Journal of Geophysical Research* **64**(4), 457–459.

Haitjema, H. M. (1995), *Analytic element modeling of groundwater flow*, Academic Press.

Haitjema, H. M. and Mitchell-Bruker, S. (2005), 'Are water tables a subdued replica of the topography?', *Groundwater* **43**(6), 781–786.

Hantush, M. S. and Jacob, C. E. (1955), 'Non-steady radial flow in an infinite leaky aquifer', *Eos, Transactions American Geophysical Union* **36**(1), 95–100.

Harr, J. (1996), *A civil action*, Vintage.

Harr, M. E. (1962), *Groundwater and seepage*, McGraw-Hill New York.

Harris, C. R., Millman, K. J., van der Walt, S. J., Gommers, R., Virtanen, P., Cournapeau, D., Wieser, E., Taylor, J., Berg, S., Smith, N. J., Kern, R., Picus, M., Hoyer, S., van Kerkwijk, M. H., Brett, M., Haldane, A., del Río, J. F., Wiebe, M., Peterson, P., Gérard-Marchant, P., Sheppard, K., Reddy, T., Weckesser, W., Abbasi, H., Gohlke, C. and Oliphant, T. E. (2020), 'Array programming with NumPy', *Nature* **585**(7825), 357–362.
URL: *https://doi.org/10.1038/s41586-020-2649-2*

Hunter, J. D. (2007), 'Matplotlib: A 2d graphics environment', *Computing in Science & Engineering* **9**(3), 90–95.

Jacob, C. E. (1950), *Flow of groundwater*, John Wiley, New York, chapter V, pp. 321–386.

Johansson, F. et al. (2013), *mpmath: a Python library for arbitrary-precision floating-point arithmetic (version 0.18)*. `http://mpmath.org/`.

Kluyver, T., Ragan-Kelley, B., Pérez, F., Granger, B. E., Bussonnier, M., Frederic, J., Kelley, K., Hamrick, J. B., Grout, J., Corlay, S. et al. (2016), Jupyter notebooks?a publishing format for reproducible computational workflows., *in* F. Loizides and B. Schmidt, eds, 'Positioning and Power in Academic Publishing: Players, Agents and Agendas', IOS Press, pp. 87–90.

Kruseman, G. P. and De Ridder, N. A. (1990), *Analysis and evaluation of pumping test data, ILRI publication 47*, International Institute for Land Reclamation and Improvement, The Netherlands.

Lutz, M. (2013), *Learning Python*, 5th edn, O'Reilly.

Polubarinova-Kochina, P. (1962), *Theory of ground water movement*, Princeton University Press.

Post, V. E., Houben, G. J. and van Engelen, J. (2018), 'What is the Ghijben-Herzberg principle and who formulated it?', *Hydrogeology Journal* **26**, 1801–1807.

Stehfest, H. (1970), 'Algorithm 368: Numerical inversion of Laplace transforms [d5]', *Commun. ACM* **13**(1), 4749.

Strack, O. D. L. (1976), 'A single-potential solution for regional interface problems in coastal aquifers', *Water Resources Research* **12**(6), 1165–1174.

Strack, O. D. L. (1984), 'Three-dimensional streamlines in Dupuit-Forchheimer models', *Water Resources Research* **20**(7), 812–822.

Strack, O. D. L. (1989), *Groundwater mechanics*, Prentice Hall.

Strack, O. D. L. (2003), 'Theory and applications of the analytic element method', *Reviews of Geophysics* **41**(2).

Strack, O. D. L. (2017), *Analytical groundwater mechanics*, Cambridge University Press.

Theis, C. V. (1935), 'The relation between the lowering of the piezometric surface and the rate and duration of discharge of a well using ground-water storage', *Eos, Transactions American Geophysical Union* **16**(2), 519–524.

Thiem, G. (1906), *Hydrologische methoden*, J. M. Gebhardt, Leipzig.

Toth, J. (1963), 'A theoretical analysis of groundwater flow in small drainage basins', *Journal of geophysical research* **68**(16), 4795–4812.

USBR (1995), *Ground water manual*, U.S. Dept. of Interior, Bureau of Reclamation, Washington, D.C.

Verruijt, A. (1970), *Theory of groundwater flow*, Macmillan civil engineering hydraulics, Macmillan.

Virtanen, P., Gommers, R., Oliphant, T. E., Haberland, M., Reddy, T., Cournapeau, D., Burovski, E., Peterson, P., Weckesser, W., Bright, J., van der Walt, S. J., Brett, M., Wilson, J., Millman, K. J., Mayorov, N., Nelson, A. R. J., Jones, E., Kern, R., Larson, E., Carey, C. J., Polat, ., Feng, Y., Moore, E. W., VanderPlas, J., Laxalde, D., Perktold, J., Cimrman, R., Henriksen, I., Quintero, E. A., Harris, C. R., Archibald, A. M., Ribeiro, A. H., Pedregosa, F., van Mulbregt, P. and SciPy 1.0 Contributors (2020), 'SciPy 1.0: Fundamental Algorithms for Scientific Computing in Python', *Nature Methods* **17**, 261–272.

Vogel, J. (1967), Investigation of groundwater flow with radiocarbon, *in* 'Isotopes in hydrology. Proceedings of a symposium', International Atomic Energy Agency (IAEA), pp. 355–569.

Index

analytic elements, 137
anisotropy, 6, 172
aquifer, 4
  confined, 4
  semi-confined, 25
  unconfined, 4
aquifer test, 156
aquitard, 4
area-sink, 148
areal recharge, 10

background flow, 117
block response, 82
Boussinesq equation, 92
branch cut, 104, 126
brentq, 131

capture zone, 122
capture zone envelope, 118
complex coordinate, 125
complex potential, 125
complex variable, 125, 201, 206
complimentary error function, 71
conformal mapping, 175
continuity of flow, 15
contour, 212
contour labels, 212
critical discharge, 127, 131

Darcy's law, 2, 6
density of freshwater, 55, 56
density of saltwater, 55
density of water, 3, 55
discharge of well, 99
discharge potential, 43
  for confined flow, 51
  for confined interface flow, 57
  for unconfined flow, 43
  for unconfined interface flow, 60
discharge vector, 7
divergence, 10, 15, 97
doublet, 104
drained area, 25
drawdown, 153
Dupuit-Forchheimer approximation, 7

effective transmissivity, 24
eigenvalue, 168
elevation head, 1
Euler's constant, 112
exponential integral, 130, 152
extraction well, 104

flow net, 104
fmin, 86, 157
Fourier series, 189
fsolve, 50, 217

Ghyben–Herzberg equation, 56
Glover solution, 191

Hantush solution, 161
head gradient, 2, 6
head-specified line-sink, 146
head-specified well, 137, 142
hydraulic conductivity, 2, 3
hydraulic head, 1
hydrostatic pressure, 1, 7
hyperbolic function, 26

image well, 106
impermeable base, 4
impermeable boundary, 18, 44, 106
inhomogeneity boundary, 108
injection well, 104
interface, 55
interface flow, 55
intrinsic permeability, 3

Kruseman and de Ridder, 157

Laplace transform, 87, 163
Laplace's equation, 96, 97
leakage factor, 26
leakage factor for two-aquifer flow, 35
leaky layer, 4
leaky stream bed, 127
least squares, 85, 157
linalg.solve, 138
line-sink, 144
low-permeable layer, 25

mass balance, 9, 69
method of images, 105
modified Bessel function, 111
modified Helmholtz equation, 87
multi-aquifer system, 4

numerical integration, 75, 218
numerical integration of pathlines, 120

object-oriented programming, 138
observation well, 1, 157

partially penetrating stream, 174
pathline, 120
periodic flow, 77, 158
phreatic storage, 5, 69, 91
phreatic surface, 4
physical plane, 175
piecewise constant, 108
piezometer, 1
point-sink, 144
Poisson equation, 10, 95
polder, 25
porosity, 3
potential flow, 43
pressure, 1
pressure head, 1
pumping test, 156

quad, 75, 161, 218

radial coordinates, 96
reference plane, 175
reference point, 142, 149
representative elementary volume, 3
resistance of leaky layer, 25
resistance to vertical flow, 15
response to flood wave, 76
river bed conductance, 20

salinity, 55
saturated thickness, 43
seawater intrusion, 193
semi-confining layer, 25
skin effect, 164
solve_ivp, 120

specific discharge potential, 171
specific discharge vector, 6, 171
specific storage coefficient, 6
specific yield, 5
stagnation point, 38, 118
steady flow, 5
Stehfest algorithm, 88, 163
step response, 82
step in the base, 48, 181
storage coefficient, 5, 69
Strack potential, 56
stream function, 16, 97
    in the horizontal plane, 97
    in the vertical plane, 16
streamline, 16, 97
streamplot, 11
superposition, 75, 101, 106, 137, 152

Theis solution, 152
Thiem solution, 100, 112, 155
time series analysis, 86
tip of the interface, 55, 65, 191, 195
toe of the interface, 55, 62, 66, 134
Toth problem, 185
transient flow, 5, 69
transmissivity, 4
travel time, 12, 18, 46, 120
two-aquifer system, 35, 39, 113, 167

uniform background flow, 118
uniform flow, 117

variable saturated thickness, 43
variable transmissivity, 91
vectorize, 213
velocity, 6, 12, 18, 120
vertical exaggeration, 12
vertical plane, 15, 171
viscosity, 3
volume balance, 9, 15, 69
volume balance equation, 9

water balance, 9, 70, 75, 95, 96, 164
water divide, 14
water table, 1, 4
well, 96
wellbore storage, 164